NATURAL BEEKEEPING

NATURAL BEEKEEPING

Organic Approaches to Modern Apiculture

Ross Conrad

Chelsea Green Publishing
White River Junction · Vermont

Unless otherwise noted, drawings are by Elayne Sears.

Editor: Ben Watson
Project Manager: Collette Leonard
Copy Editor: Susannah Noel
Proofreader: Nancy Ringer
Cover Design: Peter Holm, Sterling Hill Productions
Interior Design: Robert A. Yerks, Sterling Hill Productions

Our Commitment to Green Publishing

Chelsea Green sees publishing as a tool for cultural change and ecological stewardship. We strive to align our book manufacturing practices with our editorial mission and to reduce the impact of our business enterprise on the environment. We print our books and catalogs on chlorine-free recycled paper, using soy-based inks, whenever possible. Chelsea Green is a member of the Green Press Initiative (www.greenpressinitiative.org), a nonprofit coalition of publishers, manufacturers, and authors working to protect the world's endangered forests and conserve natural resources.

Natural Beekeeping was printed on 60# Joy white, a 30 percent post-consumer-waste recycled paper supplied by Thomson-Shore, Inc.

Printed in the United States of America
First printing, June, 2007
10 9 8 7 6 5   10 11 12 13

Library of Congress Cataloging-in-Publication Data
Conrad, Ross, 1963–
  Natural beekeeping / Ross Conrad.
      p. cm.
  ISBN 978-1-933392-08-0 (pbk.)
  1. Bee culture. 2. Organic farming. I. Title.

SF523.C75 2007
638'.1—dc22

                          2007007247

Chelsea Green Publishing Company · Post Office Box 428, White River Junction, Vermont 05001
(802) 295-6300 · www.chelseagreen.com

Disclaimer
The use of the hive management techniques described in this book is strictly your responsibility. Anything that is used with the intent to kill unwanted organisms (pests or weeds) is considered a pesticide by the United States Environmental Protection Agency (EPA) and therefore falls under its authority and jurisdiction. EPA regulations prohibit the sale of honey and other bee products taken from hives treated with unapproved pesticides, no matter how safe or nontoxic they may be. It is your responsibility to see to it that any hive products you offer for sale to the public do not violate this requirement.

DEDICATION

Dedicated to you, the Reader. May the honey bee bring as much joy and wonder into your life as it has to mine.

ACKNOWLEDGMENTS

I would like to thank the following people, who—each in their own unique way—assisted, supported, and/or encouraged this book along to its fruition:

Jeff Cunningham, Nicole Dehne, Alice Eckles, Peter Grant, Geri Hens, Dee Lusby, Larry Plesent, Suky Shoen, Kirk Webster, Hilde Whalley, and Joshua White.

Special thanks to John Barstow, Ben Watson, and all the good folks at Chelsea Green Publishing for their belief in the value of this book and their willingness to take a chance on an ambitious first-time book author.

Extra special thanks to John Cleary, Mark Conrad, C. Wallon Hazen Denney, James Gabriel, and Sandy Lincoln for their comments on various sections of the original draft manuscript.

CONTENTS

FOREWORD

Gary Paul Nabhan

In *Natural Beekeeping*, Ross Conrad takes us on a journey, from his initially casual, coincidental involvement with honey bees as part-time wage work to his multifaceted engagement with these colonial creatures as collaborators in a struggle to keep the world we live in productive, safe, and healthy. He begins, as many of us do, with barely an inkling of how the lives of domesticated bees are intertwined with our own. Gradually, as he accumulates skills and as his passion for these creatures deepens, he attempts to solve problems that honey bee colonies now face, ones that natural and agricultural scientists have somewhat neglected.

Conrad uses honey bees as the proverbial "canaries in the coal mine" to provide a sense of the health and natural wealth remaining in the world around us, a world that has been damaged by the chemical and physical fragmentation of habitats that both honey bees and humans require to survive. He ingeniously finds ways to deal with the parasites and pesticides that trip up honey bees as they go out, day after day, to do their sacred work of pollination. Ultimately, because Ross Conrad buzzes and brainstorms where other angels have feared to tread, he makes organic apiculture seem not only possible but necessary.

In many ways, this is the worst of times for the honey bee in America, as the species *Apis mellifera* is facing unprecedented threats.

There are fewer honey bees to be seen today in the wild and agrarian habitats of this continent than were witnessed by any human generation during the last 150 years. This is because of the inadvertent introductions of tracheal and varroa mites during the 1980s, the arrival of Africanized bees in our southern states (and the killing off of hives genetically contaminated by them), the prevalent and persistent use of agricultural chemicals, and, most recently, the so-called colony collapse disorder (CCD) that is decimating many managed honey bee populations. The dramatic die-offs of honey bee colonies that are occurring around the time this book is being released have not yet been fully explained by scientists, but there is little doubt that honey bee exposure to pesticides and other agricultural chemicals has weakened individual resistance and disrupted their complex of immune responses.

In short, the insights embedded in *Natural Beekeeping* would be interesting and welcome to beekeepers living at any point in history, but they are all the more vital, all the more urgent, at this peculiar moment in time. As I read Conrad's litany of the perils that the formerly ubiquitous, highly adaptable, unflappably resilient domestic honey bee faces, I couldn't help but worry even more about the tens of thousands of other truly wild pollinator species in America, whose work sometimes complements and sometimes competes with that of the domesticated colonists. Wild bees, butterflies, bats, and birds are getting slam-dunked by many of the same insults that affect the honey bee, and many are even more susceptible to pesticides, due to their body weights and behaviors. Most of the honey-bee-friendly agricultural practices that Conrad champions are, by and large, pollinator-friendly across the board. There are, of course, best practices that specifically aid the health of managed colonies of honey bees in ways that do not pertain to other wild pollinators, but they do pertain to us, in the sense that we consume bee products for food and medicine.

"Do least harm" is a tenet shared not only among Native Americans, Buddhists, and Jainists but among many others of diverse ethnicities and faiths as well. Ross Conrad has walked a path of learning how to do least possible harm to the honey bee, and emulating him may teach the rest of us how to live with this precautionary principle and make it evident in our own actions.

February 2007

Gary Paul Nabhan is the founder of the Forgotten Pollinators campaign, the Migratory Pollinators Project, and the Renewing America's Food Traditions (RAFT) consortium. He is coauthor or editor of *The Forgotten Pollinators, Conserving Migratory Pollinators and Nectar Corridors in Western North America,* and *Coming Home to Eat.*

PREFACE

Read not to contradict nor to believe, but to weigh and consider.

—Francis Bacon

I did not develop my passionate affection for the honey bee early in life. In fact, other than the occasional sting from stepping on a bee barefooted, or enjoying some honey in my tea, my interaction with the bee world was extremely limited. This changed during the summer of 1989. After moving to Vermont in the winter of '88, I had landed a job at a radio station and was finding it difficult to make ends meet. The rate of pay at my small-town station was quite different from what I had been used to in Manhattan, and as a result, I picked up part-time work from time to time to help pay the bills. That was the summer I met Charlie Mraz.

Charles was in his eighties and still heated his home with a wood-stove, as is the fashion throughout the state of Vermont. As a result of his advanced years, Charlie hired me to assist him in stacking the four cords of firewood he had purchased for the upcoming winter. After the work was completed, I asked Charlie if he had anything else I could help him with. He said he didn't, but that he would speak with his son Bill, who might have some work over at the honey house for me.

Charlie had moved to Vermont in the late 1920s and had started a beekeeping business called Champlain Valley Apiaries. He had gone on to build the company up into the largest beekeeping outfit in Vermont that kept all its bees within the state. Although Charles was still very involved in the business, some sixty years after its formation, his son Bill had taken over most of the day-to-day activities of running the company. I helped harvest and extract the honey that the bees produced for Champlain Valley Apiaries that summer. After this experience, I started my own disc jockey service and refocused my attention on the broadcast/music industry.

During the winter of 1990–91, I found myself working as the DJ at a nightclub near the Killington ski resort. I was working about six hours every night and had my days available to pursue any interest or activity that happened to strike my fancy. In fact, this was the first time since I had been a toddler that I had my days totally to myself. As a result I found myself doing a lot of thinking, reflecting, and soul-searching. I began to seriously evaluate my life. I also began to read more and was drawn to a book written by a Native American Indian elder named Sun Bear. Sun Bear had founded the Bear Tribe, which sought to teach Native wisdom to anyone who was willing to learn about the old ways that had been handed down to him by his parents and grandparents. This book resonated with what I was feeling at the time, and as a result, I decided to spend a month with the Bear Tribe at their teaching center known as Vision Mountain near Spokane, Washington.

It was from Sun Bear and the Bear Tribe that I received a new appreciation for nature. I began to see how little I had respected and cared for the natural world around me, and I took a vow, to myself, to bring more healing to the earth and, by extension, those around me. While I was with the Bear Tribe, I also had the fortunate opportunity to participate in several American Indian ceremonies. One of these ceremonies was a vision quest. This ancient mystical journey is undertaken as an odyssey of self-knowledge and self-fulfillment—a spiritual journey into the wilderness and the soul. Traditionally boys undertook the vision quest as they made the transition from childhood to manhood. As part of my quest, I sat alone for four days and four nights, fasting, praying, and seeking vision on a mountain where the Bear Tribe's learning center was located. It was during this time, on my second day on the mountain, that a single bee came to visit me. I looked down at one point, and there on my big toe sat a honey bee.

How it got there, I do not know. I didn't see it arrive. Nor had I felt its presence on my skin.

Native wisdom teaches that we can learn from everything in nature, so I spoke to the bee and let it know that I was open to learning whatever it had to teach me. The bee then spent about five minutes flying around me, landing occasionally on various parts of my body, before flying off on its merry way. I didn't think anything more of the bee's visit until the next day while sitting in the same spot, when a bee flew by. It came up to my ear and hovered there for a moment, buzzing, and then it flew over my head and buzzed in my other ear before taking straight off to my right, as if it had been on a beeline and my head was in its way. The bee's flight path would have looked like one of those fake arrows that comedians put on their heads to make it look like they have been shot. Whether it was the same bee that visited me the previous day I do not know for sure, but I suspect that it was. It almost seemed as if the bee were trying to speak to me and say, "Now don't forget what I told you yesterday."

I returned to Vermont following my time with the Bear Tribe. I didn't think much more about honey bees until about six months after my return, when Bill Mraz stopped by to see me. This was the first time I had seen him in quite some time. He told me that his primary beekeeping assistant had been badly injured in a car accident and would be unavailable for the remainder of the season. He wondered if I was interested in filling the position on a full-time basis. After some thought, I accepted the offer and spent the next six years receiving a wonderful education in a whole new area.

Not only did I find the honey bees fascinating, but I also became intrigued by the work that Charlie was doing with apitherapy. I was astounded by how effective the various products of the beehive were in restoring health and vitality to individuals with diseases or other health conditions. Charlie had drawers full of letters from people with debilitating diseases who had found relief by using honey, pollen, propolis, or bee venom therapeutically. A typical letter told of a person so crippled she had been confined to a wheelchair for years. Doctors were unable to do much more than relieve her pain and make her comfortable. After apitherapy treatments, she was up and walking around as if she had never been sick in the first place. These stories fascinated me and led me to research how apitherapy worked and why. Curiosity got a firm hold of me, and I began looking into other forms of alternative healing such as homeopathy, herbs, natural foods, and dietary supplements.

All these experiences have nurtured a connection between the honey bee and myself that is deeper and more meaningful than I can put into words. The bees have taught me so much over the years and have brought so many wonderful people into my world. It is my hope that I will always have honey bees around me, until the end of my days.

Since the honey bee exerts such a powerfully beneficial influence on the natural world around us, it seems logical to assume that our own efforts to help the honey bee thrive can indirectly benefit all of nature. As a result, we beekeepers are, for the most part, a collegial lot; we exist in a kind of friendly competition with one another. We certainly do not have a situation where there are too many beekeepers in the United States. In fact, I believe that we need more beekeepers in this world, and those beekeepers need to be successful in apiculture. This motivates me to do whatever I can to assist my fellow apiculturists, which includes answering questions, offering advice, and even occasionally providing on-site inspections and evaluations of hives. The most common questions I encounter revolve around how I manage to keep my bees healthy and achieve relatively low winter losses compared to the conventional beekeepers all around our area, who typically experience much larger losses. By sharing what I've learned with less experienced folks, I hope to aid them in keeping their bees alive and healthy. I also hope that, through the good work we accomplish with our honey bees, we can all help to leave the world in better condition than we've found it. In these pages, I share some of the insights I have gained, along with specific "how to" information on keeping bees in an ecological way, in the hope that others will find something of use that may be applied to their work in apiculture. This is my primary motivation for writing this book. As with beekeeping in general, it certainly isn't for the money.

During my own efforts over the years to obtain information on natural, nontoxic approaches to beekeeping, I have noticed that there are very few sources that list, all in one place, a critical mass of information on the numerous organic techniques that can utilized by beekeepers. It is my hope that the information contained in this book will concentrate a significant number of natural and sustainable solutions in a practical way, to give both the commercial and the hobby beekeeper a variety of ideas and suggestions for reducing chemical usage and to provide a number of alternative options in the effort to keep bees healthy without toxic chemotherapy controls. The more options we as beekeepers have at our disposal, the greater the opportunity for flexibility, something that is sorely needed today to help us adapt successfully to changes and events that challenge our beekeeping operations.

I'll be the first to tell you that I do not have all the answers. In fact, no small part of my enjoyment of the craft of beekeeping stems from the creativity, adventure, and sense of discovery that is inherent in the process of attempting to maintain healthy hives in a natural and sustainable way. From the very beginning, I have always refused to use toxic chemicals in my hives. My stubbornness caused me to lose many hives in the early days, but by persevering I have proven to myself that it is possible in this day and age to keep honey bees without resorting to the use of dangerous synthetic chemical compounds. I have attempted to present my experiences and what I have heard or read in a manner that will be of the most benefit to others in regard to natural and organic apiculture. As a result, I have taken pains to indicate which techniques I have directly tried and experienced and which I have only heard or read about in passing.

I see the activity of beekeeping in much the same way that I view activities like gardening and raising children. There is no single correct approach that applies to everyone. Each one of us who participates in the craft of apiculture will develop approaches and techniques that are unique to our particular needs, styles, and situations. As with gardening and child rearing, how individuals approach beekeeping will depend on their resources, knowledge, purpose, level of confidence, and philosophies for becoming involved in the activity. A full-time migratory beekeeper, for example, will manage hives quite differently from a commercial honey producer. And both of these approaches to beekeeping will differ greatly from those of a person who maintains a hive simply as a hobby in order to provide themselves with a source of honey bees for apitherapy. No matter our reasons for keeping bees, it is beneficial to seek out beekeeping information from a variety of sources and to choose those forms of hive management that best fit our personal situation, goals, and finances.

Thanks to the continual development of innovative technologies and management techniques, new methods seem to surface every year that fit the natural and organic philosophy, and improvements are gradually being made to many of the older ways of going about things. Some people may choose to incorporate just a few of these ideas into their hive management routines and simply reduce the number of chemical treatments they need to apply on a yearly basis. This would lead to an integrated pest management (IPM) approach, which dictates that treatments are applied not simply routinely but only according to need. Others may choose to be more aggressive in implementing non-toxic beekeeping and manage their hives in an entirely natural manner, as I endeavor to do. Either way, significant cost savings can be gained from reduced chemical expenditures. More important, greater peace of mind will be obtained, by easing concerns about toxic chemical contamination and exposure. The final outcome—healthier bees and robust hives—is everyone's ultimate prize.

NATURAL BEEKEEPING

CHAPTER ONE

WHY ORGANIC BEEKEEPING?

We must be the change we wish to see in the world.

—Mahatma Gandhi

The Hive as Teacher

We can learn so much from the honey bee, who also goes by the Latin name of *Apis mellifera*. Of all the insects, the honey bee seems to lend itself most perfectly to anthropomorphism. For example, the relationship between the bee and the plant kingdom is a powerful and intricate orchestration of interdependence and cooperation. To live its day-to-day life, the bee need only collect nectar and pollen from the flowers in bloom. These gifts from the plant kingdom, along with some water, plant resins that the bees use to make propolis, fresh air, and sunshine, are all the bees need from the world around them to survive and prosper within their colony. Thus, unless it feels threatened and is forced to defend itself or its hive, the bee is the only creature in the animal kingdom, that I am aware of, that does not kill or injure any other being as it goes through its regular life cycle. *Apis mellifera* damages not so much as a leaf. In fact, honey bees take what they need in such a way that the world around them is improved. By pollinating blossoms during nectar- and pollen-foraging activities, the honey bee contributes directly to the abundance found on Earth. This industrious little creature even transforms the nectar it collects from sugar water into deliciously sweet and health-promoting honey.

As a result, beekeeping is a wonderful way to give back to the world and help make it a better place, while at the same time receiving many incredible gifts. It is a pleasure to know that, although they may not realize it, my neighbors down the road are able to enjoy a bountiful harvest from their garden due, in part, to the pollination services rendered by the honey bees in my care. I like the fact that my hives, working in conjunction with the local plants, help produce an abundance of nuts, seeds, fruits, berries, et cetera, so the birds and other wildlife living in the area around my bee yards have plenty to eat. Because many of the seeds and nuts that are not eaten will develop into a variety of beautiful flowers, herbs, bushes, and trees, my neck of the woods is more likely to be a beautiful and bountiful place to live for many years after my passing. By directly participating in the creation of numerous seeds, nuts, fruits, and vegetables through cross-pollination, the honey bee benefits plants, animals, and humans alike. Farmers know that as the number of bees in an area increases, so will the quality of the fruits and vegetables grown in the region.

All the while, the bee is going about the business of creating honey, a substance that not only tastes wonderful but can help heal burns, cuts, infections, and numerous other maladies of the body. So much has been written about the amazing healing aspects of honey that I will not go much further into this topic other than to encourage the reader to investigate the many benefits of this "first-aid kit in a jar." Honey is something so precious and special; even with our highly developed technological sciences, we humans still have not been able to duplicate the efforts of the simple honey bee and create the same substance from what amounts to nothing more than sugar water.

The honey bee inspires me to work into my daily life this lesson: that we should take what we need to live in the world in such a way that we give something back and improve upon things, thus making the world a better place. Many of the world's problems could potentially be solved in short order if every person took this single lesson from the honey bee to heart and worked to manifest it in his or her life.

The indigenous peoples that inhabited the Americas prior to the arrival of Europeans shared a philosophy that regarded the natural world as a teacher. They saw that in the natural balance of life, extremely diverse organisms live in coexistence. Everything has its place in the order of things. Unfortunately, this tolerance seems to have been lost in the industrial model developed by our Western culture. Although the industrialization of agriculture offers much promise, it also has created many serious problems. This industrial model produces the zero-tolerance mentality that gives birth to the toxic chemical treatments typically used in agriculture, and recently embraced by many in the beekeeping community.

The propensity of the honey bee to focus on one type of flower during each foraging trip makes it ideally suited for crop pollination. Photo by Steve Parise.

A more holistic approach, one that views the honey bee and the hive environment in terms of a biological model, stems from a nature-based beekeeping philosophy. In this model, the concepts of coexistence and tolerance are included. This perspective precludes the use of toxic chemical solutions that seek to wipe out every pest in the hive. The holistic viewpoint recognizes the unique role played by all of the creatures of creation. Everything has its place and reason for being, even if it is not immediately obvious. I must admit to not being privy to the reason for the existence of the mite known as varroa (*Varroa destructor*), which has become the bane of the beekeeping industry in many parts of the world. However, I am certainly thankful for its presence, if for no other reason than the trouble it has caused beekeepers and honey bees has inspired so many people to seek alternatives to the reigning beekeeping paradigm of our times: the industrial model. Once again, nature is teaching us to seek out a bigger picture.

Toxic Chemicals Infiltrate the Classroom

As of 1987, the keeping of honey bees was the only widespread agricultural endeavor in the United States that had not become reliant on toxic chemicals to secure a harvest. In that year the parasitic mite

Varroa destructor, formerly *Varroa jacobsoni,* was first observed in hives located in the United States. This mite—known to infest honey bee colonies and suck blood from the bees, causing weight loss and deformities, spreading disease, suppressing immunity, and reducing life span—has since spread throughout America, assisted in part by the unwitting cooperation of migratory beekeepers. The overwhelming hive losses experienced by U.S. beekeepers in subsequent years prompted an outcry from the beekeeping industry for assistance. The response followed the same path tread by other agricultural commodity groups that have had a need to control insect pests threatening their crops. Beekeepers turned to toxic chemical compounds to solve their mite problem. Unfortunately, hindsight shows that this approach is a short-term solution, at best.

Since the end of World War II, when pesticides first became widely utilized, insects have consistently developed resistance to the toxic chemicals that have been used to control them. The response from agribusiness has been to turn either to higher doses of chemicals or to more toxic compounds in an effort to keep pests in check. History repeated itself with the beekeeping industry's approach of inserting chemical-impregnated plastic strips into the hive to control varroa mites.

The first strip was released in 1992 and went by the brand name Miticur. The strip, when introduced into the hive, carried with it a dose of the chemical amitraz, which was designed to be toxic enough to kill the mites, but not so potent as to kill the bees on which the mites reside. This approach to pest control is similar to using chemotherapy drugs in the treatment of cancer. Chemotherapy drugs are toxic to healthy cells as well as cancer cells. However, the chemotherapy treatment protocol calls for a dosage that is toxic enough to kill the cancer before the drugs have the effect of killing the person hosting the disease. Unfortunately, many beekeepers continued to experience huge hive losses despite treating with Miticur, so it was removed from the U.S. market after a short duration, although it continues to be marketed in Europe under the trade name Apivar. At the time of its discontinuation, some people believed there had been a problem with the strips containing the wrong dosage of pesticide, which failed to kill the mites out-right and allowed the mite to develop resistance to the chemical. However, our lack of knowledge and experience with this new mite may have also played a role in Miticur's apparent failure. Unless the unlikely event of production problems during pesticide manufacturing occurred, the beekeepers who initially experienced major losses to the mite despite their use of Miticur were probably treating their hives too late in the season, without realizing it. Remember, this was in the early 1990s, and beekeepers had not yet gained much experience dealing with the varroa threat.

The next pesticide strip to be utilized in the hive was Apistan, which relied on the active chemical component fluvalinate. Fluvalinate is related to the pyrethrin family of pesticides that are synthesized from plants and is unlike its more toxic and long-lasting cousins, that are derived synthetically from petroleum. In contrast to Miticur, this product initially gained success on the market. Unfortunately, after several years of Apistan use, resistance to fluvalinate became widespread among varroa mites, and the industry turned to a much more toxic synthetic chemical compound called coumaphos. Classified as an organophosphate, coumaphos is sold under the trade name CheckMite.

Organophosphates are among the most toxic chemical compounds used in agriculture. In fact, organophosphates are so toxic and persistent in the environment that at the present time there is an effort by the U. S. Environmental Protection Agency (EPA) to wean agriculture off these compounds, with the goal of eventually removing

Australia is the only continent with bees that does not support a population of Varroa destructor . . . *so far.*
Photo by Steve Parise.

them from the market altogether. CheckMite is similar to Apistan in that it relies on a plastic strip inserted into the hive to introduce the chemical control. Unfortunately, it took only a few years before varroa began showing resistance to coumaphos as well. The conventional approach has since evolved into using a rotation between Apistan and CheckMite, altering their use from season to season in the hope that mites resistant to one chemical will not be resistant to the other. Undoubtedly, new synthetic toxic chemicals will be approved for use against varroa in the future in order to help further address the resistance factor that varroa has developed to existing treatments. Walking this chemical treadmill may prove effective in controlling varroa for a short period, until the mites become resistant to them all. Unfortunately, these toxic compounds have the ability to accumulate in beeswax and honey, and this increases the potential for chemical contamination, resulting in undesirable consequences for both beekeepers and consumers. Anecdotal stories of beekeepers using the active ingredients fluvalinate, amitraz, and coumaphos in unregistered and illegal ways further exacerbate this situation. The fact that varroa mites developed resistance to amitraz, despite the relatively short time it was available for legal use by beekeepers, supports such anecdotal evidence.[1] Increased costs associated with these "chemotherapy" treatments have reduced the profitability of the beekeeping industry. In addition, the backlash that may result if honey on the market is found to be contaminated with one or more of these pesticides, combined with their lack of effectiveness, greatly decreases the value of the chemical approach to mite control for beekeepers.

> The effect of exposing honey bees to sublethal doses of certain chemicals is rarely discussed. Just as with people, the long-term health and vitality of the hive is likely to be compromised from such exposure, even if such detrimental effects are not readily and immediately evident.

While the conscientious apiculturist will follow the directions on the pesticide label and wear gloves during application, exposure during use can still occur. Chemical contamination is most likely to result from secondary exposure, after the beekeeper touches objects with the chemically laden gloves worn while applying the pesticide. This is inevitable if the beekeeper works the bees without assistance, as is the case with most hobby and small beekeeping operations. I am reminded of the time I was assisting with the installation of some CheckMite strips in a bee yard. One person's job was to smoke the bees and remove and then replace the hive covers, while the other person, wearing gloves, inserted the strips into the hive once the covers were removed. On this occasion, while the strips were being placed into a hive, a bee flew up and stung me on my eyelid. My initial reaction of shock and pain triggered my kindhearted coworker to want to help me. He instinctively reached up to remove the barbed weapon from my face as I was groping blindly at my pulsating eyelid and having difficulty removing it myself. The look of utter horror that crossed my face as I backed slowly away from the coumaphos-laden glove caused us to break out in hysterical laughter. This was undoubtedly the fastest recovery from being stung in a sensitive area that I had ever experienced. It also demonstrates how situations can arise when even a cautious individual can be exposed to the dangerous chemical pesticides used in conventional apiculture.

As a boy, I remember hearing it reported by the U.S. Centers for Disease Control (CDC) that almost one out of every three American citizens could expect to get cancer at some point in their lifetime. Today, the CDC tells us that on average over 44 percent of the people in the United States can expect to get some form of cancer during their life, and this statistic keeps rising each year.[2] Meanwhile, the inventory of other degenerative metabolic diseases that plague our modern society continues to stack up like cordwood and is far too long to recite fully. They include not only cancer but attention deficit disorders, mental retardation, leukemia, male sterility, and birth defects—all of which can be linked to the poisons used in and around our food supply. It is said that, despite the many hazards of Western culture, beekeeping ranks high among the professions whose participants tend to live long and healthy lives when compared to people who make other career choices. It will be interesting to see if the potential for contamination due to the relatively recent advent of toxic chemical use within the beekeeping industry will adversely affect this favorable designation.

The effect of contaminated honey on consumers and their purchasing habits also becomes an issue when chemicals become an integral part of the beekeeper's management of pests. Research in Europe has shown that the chemicals used to control mites in hives are readily absorbed by beeswax. These pesticides and their chemical breakdown products have the potential to migrate into the honey that is stored in the wax combs. By approving the use of these miticides, the EPA has deemed the potential for such low-level contamination acceptable for both fluvalinate and coumaphos, claiming that allowable residue levels in honey and wax are not exceeded when these materials are applied according to the label instructions. However, as health food advocates have been pointing out for years, just because you can make it to the door after consuming minute quantities of these toxic compounds does not mean that they are benign and have no deleterious effects in the long run.

The effect of exposing honey bees to sublethal doses of these compounds is rarely discussed. Just as with people, the long-term health and vitality of the hive is likely to be compromised from such exposure, even if such detrimental effects are not readily and immediately evident. Because long-term tests have not yet been done, we may yet find that physiological and metabolic changes occur within honey bee populations over extended periods, and these detrimental effects may become obvious only after several decades of exposure. Research indicating reduced sperm counts in drones that have been exposed to coumaphos, and the negative effects on queens reared in cells constructed out of coumaphos-impregnated wax, validates these concerns.[3]

The effects of repeated exposure of bees and humans to small amounts of several toxic chemicals simultaneously is also unknown. The EPA requires safety studies be carried out only on individual pesticides in isolation, rather than in combination with one another. In actual practice, more than one chemical may be utilized by a beekeeper during the course of the season. Knowing that compounds may combine synergistically and have a greater effect when acting together than when used alone raises serious concerns about the effectiveness of the regulatory guidelines that are designed to safeguard our health and the health of our bees.

The Meaning of Organic

Despite widespread belief to the contrary, the term *organic* does not mean that the final crop or product is totally free from toxic chemical contaminants. This mistaken perception has taken hold primarily due to the efforts of manufacturers and marketers, who have successfully promoted the notion that organic products are pure and chemical free. Meanwhile, because of the success of this viewpoint, resentment has built up among many beekeepers who feel that the organic label relegates their own commercially produced honey to second-class status. What was once considered a natural, healthy product is now deemed inferior, when, in fact, the final products of conventional and organic production may not be all that dissimilar in terms of their chemical composition.

At its inception, the organic approach traditionally referred to a management style and philosophy that is biological in nature. Rather than being a statement about product purity, *organic* was all about the big picture. It referred to approaches that care for the life in the soil and minimize the use of nonrenewable inputs and energy sources, such as those derived from petroleum. Organic principles embraced an attitude of fairness and care in regard to our common environment, as well as social concerns such as the welfare of farm workers. One of the original aims of organic agriculture was to establish a sense of stewardship for the earth, embracing human-scale operations that fit harmoniously with the landscape and local community. Although it was certainly possible that organic management practices would result in a cleaner product, it was not the primary focus. Instead, organic management sought to mimic the natural world in its efforts to be sustainable, with the ability to be carried on indefinitely, as nature has proven herself to be. For example, organic farmers have long relied on beneficial insects that feed on pests in order to reduce crop damage, such as the large-scale release of ladybugs to reduce the level of aphids on a crop. Some farmers and growers use traps on a regular basis. These traps mimic the effect of natural predators by removing unwanted insects, often by luring the unsuspecting victim using a synthetic version of a natural pheromone attractant. These are just a couple of the approaches to pest control that copy the ways the natural world will spare some plants over others from the damaging effects of insect predation. Although these types of approaches may have little impact on the quality of the final crop, they are integral to the organic philosophy that stands behind the final product.

> I like the fact that my hives, working in conjunction with the local plants, help produce an abundance of nuts, seeds, and berries so the birds and other wildlife living in the area around my bee yards have plenty to eat.

In contrast, one of the guiding principles of the industrial model that has been so aptly developed by Western culture is the desire to maximize production. When applied to agriculture, this typically results in the drive to push biological organisms to the limits of their capacity. Unfortunately, the focus on increasing our harvest seems to distract our attention from the quality of the crop that is being produced and the health of the plants or livestock that are doing the producing. In the dairy industry, for example, the cow that historically produced 15 to 20 pounds of milk a day and lived for fourteen years or more in a healthy, relatively disease-free state has today been bred to pump out an average of 90 to 120 pounds of milk a day and has to be sent to the slaughterhouse within three or four years simply because she becomes exceptionally prone to sickness and disease from the stress of the forced increase in milk production. The poor cows are literally worn out. The humble honey bee is similarly affected by our efforts to artificially boost the size of its honey crop. Activities such as the use of chemical mite controls or the feeding of sugar syrups and pollen substitutes, although beneficial to honey production in the short term, ultimately weaken the vitality of the hive and increase its vulnerability to diseases and pests such as varroa. As a result, such management tools should be used sparingly, if at all.

Our industrial model encourages large-scale production under the "economy of scale" argument that has been the drumbeat of U.S. schools of agriculture since the end of World War II. That is to say, "If you want to be profitable, you must grow larger." Although this makes sense in many industries, the fallacy of this approach when applied to farming—an inherently biological activity—is spelled out by Brian Halweil of the Worldwatch Institute in the book *Eat Here: Reclaiming Homegrown Pleasures in a Global Supermarket*.

> In the post-war period, along with increasing mechanization, there was an increasing tendency to "outsource" pieces of the work that the farmers had previously done themselves—from producing their own fertilizer to cleaning and packaging their harvest. That outsourcing, which may have seemed like a welcome convenience at the time, eventually boomeranged: at first it enabled the farmer to increase output, and thus profits, but when all the other farmers were doing it too, crop prices began to fall.
>
> Before long, the processing and packaging businesses were adding more economic value to the purchased product than the farmer, and it was those businesses that became the dominant players in the food industry. Instead of farmers outsourcing to contractors, it became a matter of large food processors buying raw materials from farmers, on the processors' terms. Today most of the money is in the work the farmer no longer does—or even controls. Tractor makers, agrochemical firms, seed companies, food processors, and

supermarkets take most of what is spent on food, leaving the farmer less than 10 cents of the typical food dollar. (As noted earlier, an American who buys a loaf of bread is paying about as much for the wrapper as for the wheat.)

Ironically, then, as farms became more mechanized and more "productive," a self-destructive feedback loop was set in motion: over-supply and declining crop prices cut into farmers' profits, fueling a demand for more technology aimed at making up for shrinking margins by increasing volume still more. Output increased dramatically, but expenses (for tractors, combines, fertilizer, and seed) also ballooned—while the commodity prices stagnated or declined. Even as they were looking more and more modernized, the farmers were becoming less and less the masters of their own domain. On the typical Iowa farm, the farmer's profit margin has dropped from 35 percent in 1950 to 9 percent today. To generate the same income (assuming stable yields and prices), the farm would need to be roughly four times as large today as in 1950—or the farmer would need to get a night job. And that's precisely what we've seen in most industrialized nations: fewer farmers on bigger tracts of land producing a greater share of the food supply. The farmer with declining margins buys out his neighbor and expands or risks being cannibalized himself.

There is an alternative to this huge scaling up, which is to buck the trend and bring some of the input-supplying and post-harvest processing—and the related profits—back onto the farm. But more self-sufficient farming would be highly unpopular with the industries that now make lucrative profits from inputs and processing. And since these industries have much more political clout than the farmers do, there is little support for rescuing farmers from their increasingly servile condition—and the idea has been largely forgotten. Farmers continue to get the message that the only way to succeed is to get big.[4]

The same pressure for farmers to increase in size applies to the bee-keeping industry. Aside from the inexperienced or inattentive hobby-ist, it is the large commercial bee outfits that have had the hardest time preventing their hives from collapsing due to varroa mites. Reports of winter losses attributed to mites of 40 to 50 percent and higher are all too common among those with six hundred hives or more. In truth, the number of colonies an experienced beekeeper can manage successfully in this new era of mites is likely to be lower than those of prevarroa years, at least until new strains of bees have been developed that are either extremely resistant or outright immune to varroa.

Part of the allure of chemical pesticide use is the economic benefit that can be reaped by a reduction in labor—a cost that reflects the investment in time, and the attention to detail, required by nontoxic organic approaches to pests and disease. With the chemical approach proving itself to be less than satisfactory in many ways, those in the beekeeping industry may find themselves having to decrease their hive-to-beekeeper ratio to match their colonies with the amount of labor available to keep their hives healthy. The number of hives that a single beekeeper can inspect and treat in a timely manner is limited. This is especially true when one considers all the unexpected issues, from bad weather to flat tires, that typically arise and cause delays, forcing one to fall behind schedule. As conventional chemical treatments become less effective, the shift to nontoxic, labor-intensive management techniques will require new approaches and technologies to make up for these increased labor demands.

Aside from chemical resistance in the case of varroa mites, another key factor that affects hive survival, even when such control methods work, is the timing of mite treatments. There appears to be a certain threshold level of mite infestation within a colony that must be reached before the hive will begin to show signs of stress. An even higher threshold must be reached before the collapse of the colony is imminent. The longer the mites have to freely reproduce without hindrance, the more quickly these thresholds will be reached. By simply preventing the mite population from building up to critical levels, chances of colony survival are greatly increased, even when mites are present in the hive on a year-round basis. As a result, many beekeepers prefer to harvest honey and treat their hives earlier in the season than they used to, even though it may mean sacrificing a significant portion of the potential honey harvest.

In Vermont, this means we are usually finished taking honey off the hives by mid- to late August, so that a highly effective treatment can be applied before the consistently cool weather rolls around. Traditionally the honey harvest would not even get started before late August in our region, and often not until September, thus taking advantage of the late-blooming goldenrod plant. Goldenrod often provides a large source of nectar-bearing flowers; the autumn honey flow they create results in rich yellowish amber honey. Instead of seeing an early end to the season as a detriment, it means that the entire process of harvest and treatment can be completed prior to the start of the goldenrod honey flow. This in turn tends to result in a much lighter crop of primarily clover and alfalfa honey, which features the more delicate bouquets and flavors favored by many honey consumers. Alternatively, the late-season flows may be harvested simply by taking steps to remove some of the mites throughout the season. A key to successfully using this late-season approach is to have only so many colonies in your care that you are able to harvest and treat in time to prevent a deadly threshold of varroa mites from being reached. This can be accomplished even while honey supers, the boxes of frames specifically intended to be filled by the bees with honey and harvested by the beekeeper, are on the hive. Ongoing mite control is accomplished by utilizing physical and mechanical control measures, that will not expose the honey crop to the potential risk of chemical contamination. The consistent elimination of a percentage of mites early on will delay their population buildup and allow for the application of a traditional late-season mite treatment, without giving the mites the opportunity to overwhelm the colony.

> Reports of winter losses attributed to mites of 40 to 50 percent and higher are all too common among those beekeepers with six hundred hives or more. The number of colonies an experienced beekeeper can manage successfully in this new era of mites is likely to be lower than those of prevarroa years.

It is possible that a strain of varroa might eventually evolve that learns to keep its reproductive rate low enough so its population within the hive does not overwhelm the host colony and cause its collapse, bringing about the mites' own demise in the process. We already know that some honey bees have adapted their behavior in order to coexist with varroa. The *Apis cerana* species of Eastern honey bee, for example, has learned to live in harmony with the varroa mite over the course of hundreds, or perhaps thousands, of years.

As we shall explore, beekeeping management techniques that help to develop similar behaviors in the Western honey bee, *A. mellifera*, may lead to a climate of coexistence where the constant presence of mites may be tolerated within the hive without leading to the collapse of the colony. The sustainably minded beekeeper wishing to follow an organic approach to beekeeping should embrace this type of model. By adopting beekeeping management techniques that encourage coexistence, we humans can play an important role in the eventual development of permanent coexistence between *Apis mellifera* and *Varroa destructor.*

As a novice beekeeper, I relied greatly on the advice and suggestions of other beekeepers I met. I also read numerous books and articles on beekeeping. As a result, I am indebted to the kind and generous modern-day practitioners of this ancient craft who helped me get started on the road to becoming successful in apiculture. Today, as an active and experienced apiculturist, I find that the questions I am most often asked concern how to keep and manage honey bees without the use of toxic chemicals and still have hives survive the winters here in the northern latitudes. Perhaps this is a direct result of the increased media exposure that organic agriculture has received following the implementation of the National Organic Program (NOP), which was passed into law in 2002 by the United States Congress. It may be that the increased interest in chemical-free beekeeping that I have noticed, from hobbyists and commercial beekeepers alike, is a response to the increasing reports of environmental decline that is a result of some of our conventional agricultural techniques used on other crops. Ultimately, though, the interest in organic beekeeping may simply be a result of the use of toxic miticides and the backlash resulting from the mixed results of conventional apiculture's reliance on these chemicals to control predatory insects within the hive. Increased media reports of the ineffectiveness of conventional mite control approaches and the resulting devastating losses of hives by beekeepers have inspired a new wave of people to take up beekeeping, many of them wanting to get involved in an effort to find a way to help the honey bee survive and thrive without relying on chemotherapy treatments.

> Today, as an active and experienced apiculturist, I find that the questions I am most often asked concern how to keep and manage honey bees without the use of toxic chemicals and still have hives survive the winters here in the northern latitudes.

Some Social Implications

I believe it is important for us beekeepers to share information with the nonbeekeeping community to help educate them about the many benefits of the honey bee. All too often, about the only thing the average person knows about bees is that they sting and they produce honey. The entire beekeeping industry stands to gain from the education of the general public regarding the benefits of the bee. On a professional level, the bond that can be forged from having direct contact with members of the honey-consuming public can prove invaluable in building a local market for hive products. This form of public relations becomes more and more important as communities continue to grow and agricultural land becomes more developed and densely populated. As humans and honey bees are forced to live in closer proximity to each other, more opportunities for unfortunate encounters are likely to develop. This has already resulted in the prohibition of beekeeping activities within the confines of many municipalities throughout the United States. In the years ahead, more towns and cities are likely to ban beekeeping within their borders. The importance of positive public relations as a preventive to public ignorance and prohibition of beekeeping activities will continue to expand. Meanwhile the growing demand for, and awareness of, organic foods by the general public may provide organically inclined beekeepers with a public relations edge over their conventional counterparts. The current state of the beekeeping industry certainly points to the need for new approaches and ways of thinking.

Following the development of the atomic bomb, Albert Einstein is said to have noted, "The release of atom power has changed everything except our way of thinking . . . the solution to this problem lies in the heart of mankind. If I had only known, I should have become a watchmaker." In many ways, the current dilemma facing the honey bee mirrors the challenge faced by the human race. In my opinion, some of the most difficult yet important work each of us can be involved in is our personal growth and evolution. If we no longer want to live in a world based on fear, lies, greed, and violence, and instead want to create a world where love, truth, peace, and compassion prevail, we must start with ourselves. Each one of us has the opportunity

17

to self-evolve and play an important role in creating a more peaceful and harmonious world simply by living a life in which these values are expressed on as consistent a basis as we can possibly muster. By the same token, we cannot rely on those in positions of power and leadership to solve the myriad social and environmental issues with which we are currently faced. Just as the ultimate answer for solving the numerous difficulties facing humankind lies in the raising of our society's collective consciousness, the most desirable and permanent solution to the difficult times the honey bee is now experiencing also lies in the bee's evolutionary process, through the development of resistance to disease and parasitic pests. The goal of raising humanity's collective consciousness requires the raising of each individual's consciousness to the point where enough of us evolve to affect society as a whole. This is not something that can be forced or imposed upon individuals. It is a responsibility each of us must choose to take upon ourselves in our own time, when we are ready. So it is with the honey bee that the evolutionary process must take place one hive at a time. Just as with us humans, the bee creates its own future with each seemingly insignificant daily decision and activity.

This realization has recently led me to start the process of examining and monitoring my own thoughts, feelings, and actions on a daily basis in an effort to identify those areas of my being that I wish to improve upon. Once they are identified, I can then begin the process of analyzing the source of my motivations and make the changes and shifts in my thoughts, beliefs, and actions necessary to bring about permanent change, whether it be altering unhealthy habits, changing ways of thinking that no longer serve me, letting go of false ideas and unproductive emotional responses, and so on. Albert Einstein is but one of many who have brought forth this message that I am only now just beginning to fully comprehend and embrace in my life . . . all with the help of the bees.

Notes

1. Malcolm T. Sanford, "Using Liquid Formic Acid for Mite Control," *Bee Culture* 131 (June 2003): 18.

2. *SEER Cancer Statistics Review, 1975–2003*, table I-14, "Lifetime Risk (Percent) of Being Diagnosed with Cancer by Site, Race/Ethnicity and Sex," National Cancer Institute, http://seer.cancer.gov/csr/1975_2003/results_merged/topic_lifetime_risk.pdf.

3. Larry Connor, "More on Drone Biology," *Bee Culture* 133 (September 2005): 19–21; and Jeff Pettis, Anita M. Collins, Reg Wilbanks, and Mark Feldlaufer, "Survival and Function of Queens Reared in Beeswax Containing Coumaphos," *American Bee Journal* 146 (April 2006): 341–44.

4. Brian Halweil, *Eat Here: Reclaiming Homegrown Pleasures in a Global Supermarket* (New York: W. W. Norton & Company, Inc., 2004): 63–64.

CHAPTER 2

WORKING WITH THE HIVE

Speak with contempt of none, from slave to king,
The meanest Bee hath, and will use, a sting.

—Benjamin Franklin, *Poor Richard's Almanack*

Among the myriad things the honey bee teaches us is the value of strength in numbers. Whereas a single honey bee may prove merely irritating and bothersome, dozens of bees present a formidable and intimidating front that can be enough to drive off predators many times their size. Thus, the honey bee earns respect. Conventionally this respect expresses itself by way of a full-body bee suit that includes a veil and covers every part of the flesh, with the exception of the hands and feet, which are protected by thick gloves and sturdy shoes. While such an outfit may provide a certain level of emotional security, its physical protection all too often proves tenuous, as at least one bee always seems to eventually find an opening in which to deliver a sting, either around the cuffs, at an ankle, through a ripped seam, or through pinched material that is pressed tightly against the skin—usually in the crook of a bent elbow or knee. As a result, the beekeeper has numerous opportunities to learn to both accept and conquer the fear that shadows every move within the apiary. Although unseen, this fear is like a constant summons to the beekeeper, calling upon the inner strength and resolve that each of us possesses, bringing it forward with the confidence and calming knowledge that we can handle whatever crisis might take place, including the occasional shattering pain that stands in our path, like a sheet of glass that we must pass through as its sound echoes among the hives.

The beekeeping supply companies are not going to appreciate me for this, but from my perspective a pair of old trousers and a long-sleeved shirt combined with a veil offer as much protection as most bee suits, only with a lot less expense. Indeed, given the right approach, the veil, the shirt, and even the pants are not strictly necessary. It has been well established that stinging insects are much less likely to sting light-colored, smooth-textured clothing than clothing that is dark and rough, such as wool, as this may too closely resemble the fur of a bear or other predator, thus making the bees unruly. As a result, in choosing what to wear, lightly colored cotton clothing is the preferred attire when you are dressing for a day at the apiary.

The one exception to my bias against bee suits is found in the unique design of the coveralls sold by Golden Bee Products of Picayune, Mississippi. This suit comes complete with a built-in veil and is made of two loosely woven layers of fabric that sandwich a third layer. This center layer is a nylon mesh that acts as a spacer between the inner and outer layers of fabric. The three loose layers result in a suit that is too thick for a bee to sting through due to the space within the nylon mesh. Nor does the suit even offer many places that a bee is capable of stinging. The large number of little holes in the outer fabric creates a suit with a minimal amount of fibers that are woven tightly enough to allow a honey bee to embed a stinger. The only areas that are capable of being stung are the pocket areas, which, out of necessity, are made of fabric composed of tightly woven fibers. This airy design also provides plenty of ventilation, which makes the suit a lot more comfortable to wear in hot, humid weather than conventional suits that are made of cotton.

No matter what you wear, if you work regularly with honey bees, sooner or later you are going to get stung. Accepting this fact of life and learning how to deal with it are essential prerequisites for any apiculturist-to-be. The anatomy of a bee's stinger consists of a venom sac and muscles attached to a shaft, which is composed of two barbed lancets lying side by side, with a hollow opening down their middle that acts as the poison canal. When the stinger is embedded in the flesh, the attached muscles cause the lancets to work their way deeper into the sting site, while the venom sac pumps venom down the poison canal and into the victim. The act of pulling the stinger out with your fingers causes the venom contained within the sac to be squeezed out into the poison canal and delivered into the skin, causing greater inflammation, swelling, and irritation. Therefore it is recommended that you remove bee stingers by scraping them off the

skin with a fingernail (or the flat edge of a hive tool), rather than by grasping and pulling. By not squeezing the venom sac when removing a stinger, the amount of venom that is deposited in the site of the sting will be minimized and the associated unpleasant side effects limited.

For the roughly 1 percent of the population who suffer from hypersensitive allergic reactions, wherein the whole body responds negatively to bee stings, taking up the craft of beekeeping is not advisable. Luckily for the other 99 percent of us, this is not the case. Unfortunately, some beekeepers, and even some doctors, will overreact to the body's reaction to a bee sting. One thing I learned from Charles Mraz (1905–1999) and his use of bee venom in apitherapy is that although having one's leg swell up after being stung in the foot is certainly a severe reaction to bee venom, it is not life threatening. Nor is it the type of reaction that would warrant carrying around a dose of antihistamines and an EpiPen. A truly dangerous reaction is one where the entire body reacts, rather than just areas near the site of a sting. Symptoms such as reduced blood pressure,

Beekeeper in full bee suit. Due to its unique construction, this bee suit made by Golden Bee Products outperforms the competition when it comes to reducing the possibility of the honey bee stinging through the suit. In addition, its increased ventilation makes this suit preferable for use in hot weather. As you can see, suits age with use, and duct tape works well to seal patches where holes start to wear through. Note that the legs of the suit are tucked inside the boots to prevent the persistent bee from crawling up the pant legs.

and breaking out all over in hives are the types of hypersensitivity reactions that can lead to a fatal outcome and legitimately call for a medical antidote to be carried as a precaution.

Besides setting up the wearer for the unreasonable expectation of never getting stung, the full regalia of the beekeeper comes with additional drawbacks. Thick bee-proof gloves make it difficult to perform actions that require much dexterity. A screened veil can be a challenge to see through, especially when you are trying to view small objects, like bee eggs, nestled deep inside a wax cell. Plus all that extra gear can be very uncomfortable on those hazy, hot, and humid days of summer. Inevitably, you will be tempted to shed part, if not all, of your bee armor at some point in your beekeeping career. Luckily there are ways of doing so that can ensure adequate protection, while at the same time creating the impression to the casual observer that you are much braver than you actually may be.

Smoke

There are two primary tools that beekeeper's can use in place of some, or all, of their defensive clothing when working with honey bees. The first is the time-honored practice of using smoke to confuse and distract the residents of the hive. According to some of our most ancient historical records, smoke has always played an integral part in working with honey bees. Although the types of materials used to create smoke and the various forms of applying smoke to hives have varied throughout the centuries and from culture to culture, the basic concept has remained the same: to distract the bees and disrupt their communication. Because smoke by definition contains pollutants and toxic gases, the type of fuel used in the smoker is extremely important, and in fact it is regulated by most organic standards. Recognizing the necessity of using smoke when working with bees, modern-day regulations allow the use of only those smoker fuels that are derived from natural sources. In this context, the definition of *natural* means "in its original form, without added chemicals or ingredients." Hence, materials such as newspapers, cardboard, treated burlap, and any other material that contain petroleum-based or synthetic substances are prohibited, whereas wood shavings, pine needles, leaves, dried grasses, and sawdust are all considered acceptable.

Of all creatures in the wild, the moth is the only one famous for not having a healthy fear of fire. Most animals and insects—the honey bee included—will head the other way when they smell smoke. They do not understand fire in the same way people do. For humans, fire has become a highly relied-upon tool. We have come to appreciate the many benefits and uses of fire, from cooking our food to heating our homes, powering engines of internal combustion, and helping us to clean up and dispose of unwanted materials. Fire can be a wonderful resource; however, fire also has its wild side and must be treated with respect, for when uncontrolled it can cause massive amounts of destruction and harm. Animals are familiar primarily with fire's destructive powers, and as a result their natural instinct is to run from anything that even remotely resembles a blaze. The beekeeper capitalizes on the honey bee's natural fear of fire by sending smoke on a mission to spread alarm deep within the dark recesses of the colony. After sensing smoke within their chambers, the worker bees can often be seen gorging on honey stored within the cells of the honeycomb in preparation for a hasty retreat, because they have been misled to believe that their hive may soon be consumed by fire. When the bees are distracted by the message of imminent hive destruction, the colony's inhabitants are much less likely to worry about any intrusion into their midst. At the same time, the warning pheromones that are given off by the agitated bees guarding the entrances to the hive—and that serve to warn the rest of the colony in an effort to rally the troops to their common defense—are masked by the scent of the smoke.

Corporate America many years ago learned the benefit of limiting the ability of workers to come together and unite in defense of their mutual interest. The inability of workers to unite and unionize can lead to exploitation of the workforce by management. Fortunately the realization by many of today's employers that happy and unexploited workers are more productive and profitable in the long run has led to fairer treatment of employees by most companies. By the same token, the use of smoke to distract the bees and prevent them from organizing in force and chasing us away with their venom-tipped barbed stingers needs to be tempered by fair treatment. Unfortunately there are beekeepers who share the exploitative philosophy and who will remove too much honey, ignoring the bees' needs in favor of their own desires. These shortsighted individuals remove all the honey and simply allow the colony to die, calculating that the cost of a replacement package of bees purchased in the spring is less than the price that can be gained from the sale of the additional honey harvested. Other beekeepers will allow the bees to replace the lost honey by feeding them sugar syrup so that they are more likely to survive the winter months. These approaches show a lack of respect for the honey bee and are prohibited by many of the organic standards that are now in place.

When it comes to such an integral part of beekeeping as the use of smoke, the proper and effective use of the gaseous products of combustion cannot be overestimated. Now, some may feel that the following information is so basic that it does not require explanation. Or perhaps they think that because the smoker, the hive tool, and the veil are such basic necessities to beekeeping—incredibly common and seemingly simple in their design and use—any instruction as to their proper use is unnecessary. It is certainly true that through trial and error even a novice is able to learn many of the basics of what works and what doesn't when using these tools of the trade. Teaching yourself about beekeeping through the school of hard knocks, while not necessarily easy, can provide an excellent education—though the tuition is high, because the bees are quick to let you know when you're not making the grade.

The well-made smoker is constructed from stainless steel, will hold plenty of fuel, features a shield to prevent accidental burns, and has a stout nose that can act as a wedge to hold supers apart once they have been separated.

I have found the following approach works best for me. Keeping in mind that heat rises, clean out all the ashes and partially burned fuel in your smoker and place a little fresh, well-dried fuel in the bottom of the smoker cavity. Light the fuel and work the bellows with short, quick squeezes until the fuel is burning well. I look for flames shooting up out of the top of the smoker as a sure sign that there is "fire in the hole." It is important to be conscious of the direction of the wind and how close you, your clothing, and your equipment come to the exposed flame. You don't want to melt the screening on your veil or singe any hair, if it can be helped.

Once you have a good fire going, add another handful of fuel to the smoker, and again work the bellows with rapid, short blasts of air until this additional fuel is well lit. Now that you have a good hot fire burning in the bottom of the smoker, fill the rest of the fire chamber with more fuel, close the top, and you're ready for action. Care must be taken not to pack this final addition of fuel too tightly into the fire chamber or you run the risk of smothering the fire. Should you find that, after setting the smoker down for a period of time, the amount of smoke produced is reduced to a mere wisp, then making short, rapid squeezes of the bellows for thirty to sixty seconds should be enough to rekindle the burning embers within the smoker's fire chamber, unless the fire has gone out completely.

> Smoke has always played an integral part in working with honey bees. Although the types of materials used to create smoke and the various forms of applying smoke to hives have varied throughout the centuries and from culture to culture, the basic concept has remained the same: to distract the bees and disrupt their communication.

Once you have the smoker lit and the fire chamber full, making the most effective use of the smoker requires some practice. A smoker can be used not only to distract and calm the bees but also to push the bees in a certain direction, either toward or away from an area. A typical mistake that many beginners make is to use too much smoke, perhaps in the mistaken notion that if a little smoke is good, then a lot must be better. It takes just the right amount of smoke, at just the right time and from just the right direction, to prod the bees into going where you want them to go. Use too much smoke and the bees will just run around in circles, confused as to where to find a breath of fresh air. Use too little smoke and the bees will act as if there were no smoke at all and place their focus on things that we would prefer they would ignore, like the tip of our nose. For the bees to move, they need to be able to identify where the smoky areas are (places they will move away from) and where the areas clear of smoke exist (toward which they will naturally gravitate). The beekeeper can use this behavior to "herd" bees in the desired direction. Again, noting the direction and speed of the wind can play a critical role in either helping or hindering your efforts, and care should be taken to position oneself advantageously.

It is also important to pay attention to the quality of the smoke that issues from the smoker. The smoke should exit in thick white clouds and feel cool to the touch. A sure sign that the fire within the smoker has made its way to the top of the fuel pile and that your fuel needs replenishing is the appearance of hot, thin, gray smoke. Miss this tell-tale sign and you will likely end up with sparks, or perhaps flames shooting out of the smoker and singeing your bees. And trust me, blowing hot gray smoke, flames, and red-hot cinders onto your bees is a great way to make them upset with you really fast. If the fire is approaching the top of the fire chamber and you don't have enough fresh fuel to refill the smoker, one trick is to place a wad of green grass on top of the burning fuel. This will serve to cool the fire and will help prevent flames and sparks from exiting the smoker and getting the bees overly excited.

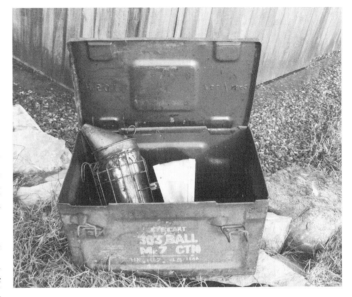

An army surplus ammunition box (minus the ammunition, of course!) provides a safe, fireproof container for storing your smoker between visits to the apiary.

When all the work is completed and the smoker is ready to be put away, it is a good idea to plug the smoke hole with a wad of green grass or a cork to limit the amount of fresh air available to the burning material inside and help smother any hot embers that remain. To be extra safe, it is wise to store your smoker in a metal can that has a lid. It can be very disconcerting to be driving down the road, look in your rearview mirror, and see flames shooting up from your bee equipment, all because a smoker was not fully extinguished before being stowed away in the back of your vehicle. It's a good idea to keep a fully charged fire extinguisher handy in any vehicle or building that is used for beekeeping purposes on a regular basis. When properly used, the smoker becomes one of the best pieces of protective beekeeping gear you can own. Just make sure you don't run out of good-quality fuel that is dry and ready to burn!

An alternative to the use of smoke as a means to make the bees manageable is the use of a one-to-one dilution of sugar in water. This solution, applied in the form of a fine spray, not only distracts the bees as they lick the sweet syrup off one another's bodies, but also serves to wet their wings down—in effect grounding them so they can't fly after you. Although not a viable option in cold weather, because it is unwise to allow the bees to become cold and wet, since they too can die from a form of hypothermia, the use of a sugar spray can come in handy when no matches or dry fuel are to be found.

Attitude

The second tool that can be utilized to help keep the bees calm is your head. In the first place, you should remember that honey bees are kind of like you and me. When it is warm and sunny, most of the foraging-age honey bees will be out and about running errands and foraging for goodies. The majority of the bees left behind in the hive will be younger individuals, many of whom have yet to develop their stinger and venom reserves. These young workers are busy storing the incoming nectar and pollen, feeding the brood, cleaning the hive, and attending to the myriad other tasks that need to be accomplished to keep everything running smoothly within the colony. However, on cool, damp, and rainy days, when it is too cold for the bees to fly and there is no nectar or pollen coming in, the hive's occupants may seem grumpy, all cooped up inside waiting around for something to do. This is not the best time to pay them a visit if you don't want to incur their wrath.

Once you are able to maintain a peaceful, focused, and present state of mind in the bee yard, the real trick is to be able to keep that same harmonious presence with you when you return to the world of humanity.

This is one of the reasons why I prefer to position my apiaries in locations that receive lots of sunshine. The extra warmth provided by the sun's golden rays could be enough, when the temperature is on the cool side, to make the difference between working with a calm hive and one with a quick temper.

Once favorable weather has arrived and you have scheduled a call on a bee yard, one of the most important things you can bring along is your peace of mind. Early on in my work with bees, I noticed that I would get stung a lot more often on days when I was irritated and not in a good mood than when I was feeling good, present, and focused on my work within the apiary. As a result, my experience has taught me that honey bees are like dogs, farm animals, and even people:

they can sense when we are calm, cool, and collected, or when we're irritable, fearful, or upset, and they respond accordingly. With every visit to the apiary, the bees reinforce within me the importance of my presence of mind. The bees will waste no time in providing me with immediate feedback when I allow myself to be grumpy, distracted, or off-balance. In their own way, they provide me with the motivation to do the work I need to do on myself in order to stay present, calm, and focused, even in situations where I could potentially be stung to death. (The lethal dose for a healthy adult ranges anywhere from about two hundred to fifteen hundred bee stings depending on body weight, age, state of health, et cetera) It is all too typical that when two or more people enter a bee yard, the person who is afraid of the bees is the first one to get stung. Once you are able to maintain a peaceful, focused, and present state of mind in the bee yard, the real trick is to be able to keep that same harmonious presence with you when you return to the world of humanity.

Protective Gear

Armed with a beautiful sunny day, a well-lit smoker, and a calm, focused demeanor, I rarely find it necessary to use a veil, let alone a bee suit, when working with bees. Now, if this is the case for me, I am sure there are others for whom these same protections work as well. I must confess, however, that I did not start out keeping bees this way. The first day that I worked with honey bees, I wore a veil and gloves, along with a long-sleeved shirt and jeans with the cuffs tucked into my socks. After all, although it is good not to be afraid of being stung, it is not something you typically want to encourage. Thus, a veil and gloves are prudent for those just starting out. However, as I slowly got used to working with the bees and became more comfortable around them, I became braver and more willing to try working without protective clothing. The first thing to go were the gloves. They were so thick and bulky that it was hard to feel anything through them, and I found that I accidentally crushed many more bees while wearing gloves than I did when I didn't have them on. I also quickly discovered that my hands were a lot less sensitive to the pain of a bee sting than many other areas of my body.

It was during a particularly hot and sweltering week working in the bee yards—with beads of sweat stinging my eyes as I viewed my more experienced coworkers from under my veil, looking much cooler and comfortable than I as they worked without similar head gear—that I first decided to try removing my veil to see what would happen. Thus began my lessons on the importance of one's attitude and mood when working in the apiary.

Readers new to beekeeping may start to think that I have been stung in the head one too many times when I state that it is incredibly beneficial to have developed enough comfort around the bees to be able to work with them on most occasions without any special protective clothing. I make this statement as a result of both personal experience and my observation of others. Now don't get me wrong. I fully recognize that a good-quality veil is handy to have around when faced with the realities of our fast-paced modern lifestyles, which may require us to perform hive manipulations under less than ideal circumstances due to our time constraints. However, the benefit of not requiring protective equipment when working with bees goes well beyond the conveniences of saving money, not having your vision impaired by a veil, and flirting with heat exhaustion inside a bee suit during the dog days of summer. When I put on a veil, I am putting up a barrier between the bees and myself, and once I have separated myself behind this protective wall it is easy to get lazy and forget to show respect for the little ladies with whom I am working. Human history has shown over and over that with separation, all too often one side will eventually start to dominate the other. Without such a division, there is greater opportunity for prolonged and peaceful cooperation.

On too many occasions, I have observed beekeepers decked out like astronauts in full protective gear carelessly crushing bees, knocking into hives, and generally handling the colonies and equipment in a rough manner. Often these are the same beekeepers who will complain that their bees are "mean" or overly aggressive, when I suspect that their bees are simply not being treated with the proper respect. It is much easier to be disrespectful when the consequences of such contempt have been largely removed. These suspicions were confirmed one day when I was asked to help a local beekeeper with his "angry" bees and I approached them as usual, with just a hive tool, lit smoker, and cap on my head to prevent the occasional curiosity seeker from getting tangled in my hair. On this occasion the bees seemed not to be aggressive at all, and I went through all the hives without getting stung once. Unless the bees have recently been disturbed, the individual working with the bees and how he or she goes about the work seem to be the primary variables that account for the differences in European honey bee behavior.

I suspect another factor in this so-called angry hive syndrome is that beekeepers making such complaints are often not cleaning their bee suits on a regular basis. As a result, their veil and coveralls are impregnated with the smell of venom from previous visits to their bee yard, and this is all that is necessary to get the bees riled up when the suited beekeeper decides to pay them a visit.

A beehive is not built by a multitude of single bees toiling away in isolation, and thus the bees have much to teach us about the value of cooperation. The honey bee acts selflessly in relation to the rest of its hive family. The vast majority of the bees' activities are done for the good of the colony as a whole, rather than to benefit any individual alone. As a result, we see that the strength of the honey bee—an insect that has survived since long before the time of the ancient Egyptians—can be traced to its ability to cooperate. This line of thought leads me to believe that Darwin's "survival of the fittest" theory, which refers to the strongest individuals as being the most likely to survive, is flawed. Closer observation indicates that those individuals who have the greatest ability to cooperate and adapt are the ones that also have the strongest likelihood of succeeding in the long run. In his later years, even Darwin himself seems to have become convinced that love (or altruism) was a stronger force than the survival instinct. Numerous examples of this phenomenon occurring in nature support his later theory, such as the behavior of packs of wolves, schools of fish, flocks of birds, and herds of buffalo. Groups of animals learned long ago that working together in cooperation greatly increases their long-term chances for survival, compared with acting alone. These are some of the thoughts that make me passionate about the craft of beekeeping— a passion that inspires me to carry on keeping bees despite the hard work, difficulties, and challenges involved. It's a passion that results in my wanting to have honey bees around for as long as possible. I would hope that everyone in the world might find something they can feel so passionate about during their lifetime.

Separating oneself from the bees with protective equipment also prevents the beekeeper from making subtle observations that may provide vital clues to problems among the hives. Without a veil, beekeepers are much more sensitive to the temperament of the colonies they work with. Not only do the bees provide instant feedback if, from their perspective, they are not being treated properly, but uncalled-for aggressiveness can point out problems that need to be addressed. To walk into an apiary and immediately have several bees angrily buzzing about your head provides an important clue that something is not right, and more investigation is needed to get to the bottom of the issue. Because of the angry bees, your heightened state of alert may call your attention to signs that may indicate there has been a dearth of nectar and no honey is coming into the hive, skunks have been feeding regularly at your apiary, or there is a drone layer amongst your hives that knows it is doomed and is letting you know that it is not pleased about the situation.

> On too many occasions, I have observed beekeepers decked out like astronauts in full protective gear carelessly crushing bees, knocking into hives, and generally handling the colonies and equipment in a rough manner. . . . It is much easier to be disrespectful when the consequences of such contempt have been largely removed.

On one occasion, while entering a bee yard located near the intersection of two dirt roads one sunny spring day, we were immediately set upon by a number of angry bees and were forced to retreat into the cab of the truck and put on veils before we could proceed with our work. This unusual reception prompted us to theorize as to the cause of such unusual activity by our typically friendly bees. While surveying the apiary, we noticed that there were a fair number of good-sized rocks that were randomly scattered among the hives and that matched, in both color and size, those alongside the dirt road that ran past this particular location. A little while later, a school bus pulled up to the corner and dropped off several children making their way home at the end of their day. The sheepish looks on the children's faces told us the story of why there were so many rocks scattered around the bee yard and why the honey bees were so jumpy.

I am pleased to report that, even though I rarely wear a veil these days, I do not get stung on a regular basis, and when I do, I find that it is typically because I am not paying proper attention to my attitude or to what I am doing and have grabbed a bee while picking up a piece of equipment or something similar. I view these occasions as reminders to pay closer attention to the task at hand, and I take solace in the idea that, typically, the more stings one receives, the more immunity one builds up to the effects of bee venom. However, despite the development of immunity to venom's effects, there seems to be no similar increase in one's tolerance for the pain involved—bee stings always hurt. If I am not careful and focused, it is likely that I may get stung several times during my visit to the bee yard. However, the good news is that it is usually only the last bee sting that hurts the most.

By leaving myself unprotected by a veil, I find that I am forced to become a better beekeeper. In addition, the knowledge that bee stings can reduce or eliminate symptoms of diseases like arthritis and multiple sclerosis helps me to accept the occasional bee sting I receive as beneficial to my health and a form of preventive health care. It's all part of the employee benefits package I receive from my job working with honey bees.

If I lived in the southern regions of the United States, it is likely that my ability to work with bees without a veil would be impossible, or at the very least ill advised, because of the Africanized honey bee (AHB). Africanized bees arose from the 1956 importation of fifty African queen bees into Brazil for research purposes. Due to the large number of natural honey bee predators that exist in Africa and the short, sporadic honey flows, the African bee developed a very aggressive defensive nature in order to protect its limited honey stores and deliver a robust response whenever its colony felt threatened. This has led to the sensationalistic moniker "killer bee" being attached by the media to this strain of honey bee, which has migrated northward over the past fifty years, to the point where it can now be found in almost all southern states in the United States, in a band that stretches across the continent from California to Florida. How much farther north Africanized honey bees will travel is anybody's guess. AHB have a difficult time surviving the winters of the Northern Hemisphere. They have a tendency to nest out in the open, leaving them little insulation and protection from the elements other than the bees themselves as they cluster. Because it is in their nature to keep the size of their colonies small and to swarm frequently, AHB are unlikely to build up a large enough population to form an adequate cluster for the winter, even if they were sufficiently sheltered; nor are they likely to store away enough honey to last throughout the cold season. That said, their maximum northward expansion will likely be augmented by the general global climate change pattern of recent years.

And yet the northern beekeeper cannot rely on the cold to keep the AHB at bay. The ease with which packages and queens can be shipped from breeders located in areas known to be colonized by Africanized bees, and the large number of migratory beekeepers that move between southern areas known to harbor Africanized bees and the northern latitudes, presents the potential risk that bees with AHB genetic traits will be shipped into northern areas. As a result, many cool-climate beekeepers are already focusing on ways to become more self-reliant in the rearing of replacement bees and queens in order to avoid the damage that a stinging "incident" is likely to cause to the image and industry of northern beekeeping. Never mind that the majority of AHB stinging episodes are nonfatal. The relatively few fatal incidents that have occurred are typically a result of hypersensitivity to bee venom, or happen when animals are tied up, or when people panic or are unable to move away from the area once the bees start to become defensive.

Another issue to consider when deciding how much protective clothing to wear, or not to wear, is the ability of some beekeepers to suddenly develop a life-threatening allergic reaction to bee venom. This is in direct contradiction to the experience of others, who have observed that repeatedly getting stung by bees allows for the development of resistance to the bee venom, so that the side effects of bee stings become less severe and eventually unnoticeable as the number of stings received increases over time. Judging from some of the accounts I have heard, the use of drugs and medications may play a role in the sudden allergic disposition to bee venom that some beekeepers develop. Certain medications can change the body chemistry of patients and reduce immune response, and I suspect that such changes may have the ability to make a person more prone to developing a life-threatening allergic reaction to bee venom when he or she has never had any allergy issues in the past.

One final but related word of caution: There are reports that members of beekeepers' families who don't work directly with honey bees themselves may nevertheless develop severe life-threatening allergies to bee stings. Some people believe that these responses are from repeated exposure to the minute amounts of bee venom the family member is exposed to from the protective clothing worn by the bee-keeping member of the family. As such, it may be advisable for bee-keepers to do their own laundry and clean their beekeeping clothing in loads that are separate from the rest of the wash.

CHAPTER 3

HIVE MANAGEMENT

A swarm in May is worth a load of hay, a swarm in June is worth a silver spoon, but a swarm in July ain't worth a fly.

—old New England beekeeper's saying

We like to believe that advancements in our society take place as old beliefs and customs slowly give way to the changes brought about by science, experimentation, education, and practical experience. However, this approach works only when the new approaches improve on old methods. In our history, agricultural advances that have boosted production all too often have overlooked long-term consequences in the rush for immediate profits and have resulted in the creation of larger, more complex problems. The world of beekeeping is no exception.

Many of the modern organic hive-management techniques that are emerging are not all that different from those of conventional apiculture. This is simply because the basics of honey bee biology and behavior must still be observed. In fact, to be successful at organic beekeeping, it is even more important to learn to work with the natural biological processes and instinctive behaviors of the bees. Whereas conventional apiculture tends to force the beekeeper's will upon the hive organism, the organic beekeeper is more inclined to work with the colony in partnership rather than in domination. As such, many of the ways of keeping bees that were common twenty to thirty years ago are being edged out for new methods and techniques that continue to change as new research and information becomes available.

One key distinction of the organic approach is that it does not hold the maximization of the honey crop as its top priority. This one seemingly small shift in perspective makes a world of difference. All of a sudden, the *quality* of the honey harvested and the health of the hive become just as important as, if not more important than, the *quantity* of honey that the beekeeper can coax out of the hive. This emphasis on quality over quantity is perhaps the defining notion of the organic agricultural movement. The attribute of quality is not limited to simple cosmetic factors such as size and looks, nor is it applied only to the resulting crop that is harvested. In organics, quality is applied to the effects of the overall operation and takes into account the health and well-being of the farm, the land, the animals, and the people involved, from farmer to consumer. Of course, quality also has to do with the taste, color, and smell of the final product. Over time, we discover that when proper attention is paid to quality, the quantity aspect takes care of itself. In the world of beekeeping this translates into the fact that, by keeping your bees alive and healthy through organic practices from year to year in the face of chemical-resistant mites, in the long term you will realize larger harvests and increased profits over those beekeepers who rely on toxic approaches that ultimately fail the bees. This can occur even though your harvest of organic honey may be less, pound for pound, than what might be obtained through conventional means, simply because the organic beekeeper is not having to spend as much money replacing hives every year and because of the premium that organic honey commands.

Apiary Location

Traditional considerations that a beekeeper needs to take into account when choosing an apiary site include not only the availability of pollen and nectar sources but other factors as well. Among these other concerns are nearby sources of clean water, yard accessibility by vehicle, good drainage, exposure to the south or east to provide adequate sunlight, windbreaks in the direction of the prevailing winds, a loca-tion that will discourage vandalism and theft, and, in highly popu-lated areas, some type of fencing or other screening so that the hives will be hidden from view and not cause neighbors to be unneces-sarily alarmed. As if all this were not enough, the organic beekeeper has additional concerns to consider when locating a bee yard, and this has become one of the most controversial aspects of organic apiculture.

Honey bees are semidomesticated, wild creatures. They will choose to forage anywhere and at any time they want, and we humans have little to say about it. As a result, bees may forage for nectar and pollen on flowers that have been raised organically; or they may visit plants that have been sprayed with chemicals or have been genetically mod-ified and therefore do not fit within the framework of organic agri-culture. As a result, current standards for organic honey often require that hives be located on a certified organic farm. Standards also require that all areas near an apiary where the bees are likely to forage must be free of activities that in any way could compromise organic integrity. Some folks have a real problem with organic regulations that address hive and foraging location and think that, as long as the hive itself is managed using natural methods that utilize nonsynthetic chem-icals, where the bees forage should be of no concern. They believe that organic standards for honey and honey bees should address only hive management and manipulations—factors that are clearly within the beekeeper's control and are thus verifiable. They reason that, even when a hive is placed on land that is managed organically, the bees are likely to fly right past the organic blossoms and forage on non-organic flowers that are within reach of the colony. A hive's foraging radius can extend significantly during times when there is a dearth of nectar. It has already been shown that about 10 percent of the honey bees in a colony may fly well over five and a half miles from the hive, and some can travel as far as eight and a half miles or more in search of food during periods when forage is scarce.[1] Additionally, if a plant in bloom has been doused with severely toxic chemicals, the foraging bee is not likely to make it back to the hive alive.

Folks who take this position are espousing what may be called a liberal viewpoint of the term *organic*. They believe in a loose read-ing of the requirements that does not follow a strict "by the book" interpretation of what *organic* has traditionally come to mean when applied to livestock. Because the biology and culture of the honey bee are significantly different from those of other farm animals, it is argued that they warrant separate regulations regarding feeding, foraging, and so forth.

In the other camp are those people who stress that the integrity of organics requires that not only must hives be managed without synthetic chemicals and antibiotics, but as much effort as possible should be made to ensure that the bees do not forage on plants that have been exposed to prohibited materials. Such areas that are considered to adversely affect organic integrity include city and town centers; industrial sites; conventional farms that grow crops that could potentially be used as a foraging source, and that have been grown with chemical fertilizers and pesticides or are genetically modified; garbage dumps and sanitary landfills; nonorganically managed golf courses; nuclear power plants; incinerators; major highway, powerline, and railroad right-of-ways; contaminated water sources; and conventionally managed apiaries. These folks take a conservative approach to the organic standards, in that they believe in a traditional interpretation of the organic philosophy when applying it to honey bees. They argue that organic standards should dictate constraints similar to those that affect other livestock, and they believe that beekeepers should follow this traditional approach to the letter. Not surprisingly, organic certification agencies tend to take a position that is more in line with this conservative approach.

> This emphasis on quality over quantity is perhaps the defining notion of the organic agricultural movement . . . In organics, quality is applied to the effects of the overall operation and takes into account the health and well-being of the farm, the land, the animals, and the people involved, from farmer to consumer.

Those who share the liberal view argue that they are being more realistic in their perspective, given that the vast majority of apiary locations within the United States are not able to comply with organic regulations as currently written, and most of the locations that do conform tend to be in areas where bee forage is extremely limited, and resulting honey crops are likely to be small or nonexistent. As a result, they point out, little certified organic honey is currently being produced within the United States due to the overly restrictive nature of the standards. Conservatives counter that it is important to meet the expectations of the organic consumer and not dilute the meaning of the term *organic* just because, unlike other livestock, bees cannot be penned in and their foraging activities controlled as easily as with cows or farmed fish, for example.

As with most arguments, each side's position has a piece that is true, but the *truth* lies somewhere in the middle. Most standards, as they are currently written, incorporate a distance limitation with respect to areas that may affect organic integrity, in an effort to find a compromise position. Requirements range anywhere from one mile to about three and a quarter miles in all directions from the apiary location, even though honey bees have been known to fly much farther under severe circumstances when searching for food. The thinking is that honey bees are like people: they are not going to work harder than they have to. If adequate nectar and pollen sources are available within a mile or so of the hive, the vast majority of foraging bees are unlikely to fly much farther afield in search of additional food. This solution thoroughly pleases neither the liberal nor the conservative points of view. If honey bees have the ability to access prohibited foraging areas just outside the regulated distances specified in organic standards, then from the conservative perspective this jeopardizes organic integrity. From the liberal viewpoint, if the bees are likely to have access to prohibited areas anyway, why bother placing any limits on apiary location at all? Doing so simply creates a false impression in the marketplace when, in these people's opinion, organic honey is not going to be all that different from conventionally produced honey.

These two conflicting points of view have yet to be fully resolved, and as a result, there is a wide range of criteria governing apiary and foraging locations that has been written into the organic standards enforced by the numerous organic certification agencies authorized by the United States Department of Agriculture (USDA). Until the USDA National Organic Program (NOP) establishes a clear set of national regulations for apiculture that will supersede all other standards, the mishmash of rules that now exist is likely to continue well into the foreseeable future, and no resolution to this controversy is likely to be forthcoming anytime soon. In fact, I suspect that it is precisely the issue of apiary location and forage availability that is responsible for much of the delay in a set of national organic regulations for apiculture being codified by the NOP.

From the honey bee's perspective, the best location is one that provides access to a wide variety of flowering plants that yield nectar and pollen throughout the growing season. Due to the fluctuations of the protein and mineral content of plants, a diversity of pollen and nectar sources helps ensure that the hive will be able to adequately provide for its nutritional needs. Access to many different types of plants will become even more important as Earth's climate changes accelerate. As various areas of Earth become warmer or cooler, the bloom cycles of plants will fluctuate. This will create more and more situations in which bees are unable to rely on the pollen- or nectar-bearing vegetation that they have relied on in the past.

An example of this are the recent reports from parts of New York State where, during an unusually mild winter, the pussy willows flowered in March instead of April. The temperatures then dropped back to seasonal norms and were too cold for the bees to fly consistently, so they were unable to collect pollen from the pussy willow blooms. This lack of access to an important source of early spring pollen adds stress to colonies. By locating hives in areas where a wide variety of plants grow, we increase the chances that colonies will be able to find adequate forage to meet their needs in spite of changing weather patterns.

Equipment

The long, cold, and dark winter months are the perfect time for puttering around in a wood shop making, assembling, and repairing the hive equipment that will see you through the upcoming honey season. There are several ways to obtain bee equipment. It can be purchased from beekeeping supply companies unassembled, or conveniently preassembled for a little extra money. Taking the time to assemble the equipment yourself gives you more control over the quality of the workmanship. You can purchase used equipment from someone else, or, if you have the tools, you can build your own from scratch. When buying used equipment one should be extremely wary of purchasing frames of drawn comb. It is within the wax of the combs that diseases and chemical residues can lie hidden. Both of these can adversely affect the quality of your honey crop and the health of your bees.

The types of materials used for constructing honey bee equipment are addressed by most organic standards, because certain materials can adversely affect the overall health and vitality of the colony or the quality of the products harvested from the hive. In the organic hive, all painting and weatherproofing are confined to the exterior of the hive, while the interior is limited to bare wood, metal, and beeswax.

Due to the presence of arsenic, conventional pressure-treated lumber is not approved for use in constructing beehives. Commercially available hive bodies and supers are typically made of pine or cypress. Not only are the types of materials used to make bee equipment important, but also the quality of the craftsmanship that goes into the construction plays a huge role in equipment usefulness and longevity of service. It is this extra attention to detail, or lack thereof, that separates the well-crafted hive from the poorly assembled one.

One key detail that some beekeeping supply companies often seem to overlook is the direction of the heartwood in relation to the exterior of the super or hive body. When a piece of lumber warps, it will bend away from the heartwood. As a result, when the heartwood is facing the inside of the hive, any warping that occurs will cause the corners of the board to pull the nails out, eventually resulting in openings being created between the boards in the corners of the hive body or super. By positioning the heartwood facing instead toward the handhold on the exterior of the equipment, the force of the board's natural tendency to bend is countered, and openings between the joints are much less likely to occur.

Another beekeeping equipment detail that could stand to be changed is the frame rest protector designed to hang down into the hive. This design seems to have become the standard industry practice. Regular cleaning is required in order to remove the wax and propolis buildup that occurs within the frame rest area. It would make more sense to design the frame rest protection so that it protects both

The well-made super features handholds on the heartwood side of the lumber.

The typical frame rest protector sold by most beekeeping supply companies provides protection to only the bottom edge of the frame rest area.

By reinforcing both faces of the frame rest area, a significant increase in the working life of the equipment can be achieved by reducing wear and tear when scraping this area clean of wax and propolis.

the horizontal and the vertical sides of the ledge rather than just the bottom edge. If the nails that fasten the metal frame rest protector in place are driven down a little below the surface, they will not catch on a hive tool when the area is being scraped clean. Frame rest protection that covers both sides of the frame rest can add many years to the life of your equipment and is well worth the investment.

One of the greatest equipment challenges faced by today's organic beekeeper is obtaining organic beeswax foundation. Beeswax acts like a sponge in that it easily absorbs and accumulates chemicals and their residues. As a result, many organic standards require the use of residue-free wax foundation, and until the infrastructure is in place to supply organic beeswax, meeting this standard will be difficult at best. Initially, any organic wax foundation that is available is likely to be both scarce and expensive. Sheets of comb-honey foundation that do not contain reinforcing wires can be cut into one-inch strips and those strips used to get the bees' comb-building process started.

If clean foundation is unavailable, beekeepers may have to drip a bead of wax along the underside of their top bars in the hopes that the bees will use it as a starting point to build natural combs without

43

foundation. Because the bees will always build their comb parallel to the force of gravity, special care must be taken to level out all hives in which a starter strip of foundation or bead of wax is being used

The thrifty beekeeper will appreciate this method of stretching the use of and thus reducing the expenses involved with purchasing foundation.

to entice the bees to build comb within the frame. If the colony is positioned at an angle, the combs built within the frames will stick out in relation to the frame's edges, creating problems during honey extraction and processing. When a hive builds natural comb neatly within a frame, the comb will be more fragile than newly drawn-out frames that utilize foundation reinforced with wires or support pins. It is advisable in the first year to extract honey from naturally built comb at a slower speed than is usually used and thus minimize damage to the combs. Typically, by the second or third year, such frames of comb have been built up and reinforced by the bees to the point where they can withstand the rigors of the rotating honey extractor without breaking, cracking, or collapsing.

While we're on the subject of frames, a mention should be made here to let the uninitiated in on a way to help keep the top bars and side bars of a frame from being pulled apart while trying to pry it loose from a hive. The trick is that after the standard two nails are hammered through the top bar down into the side bar, a third nail should be put through the side bar and into the top bar from below. This third nail is key to preventing the top and side bars from becoming easily separated.

Frames will last much longer if we take the time to add this one strategically placed nail.

Due to its simplicity, the inner cover is perhaps the easiest piece of equipment for the do-it-yourselfer to build. Unlike the majority of commercially available inner covers, which feature a raised lip around the edge on both sides and a single, oval-shaped hole in the center of the cover, I prefer covers that have a lip on one side, are flat on the other side, and feature two holes for ventilation and feeding. By positioning the cover on top of the hive with the flat side down, the colony's bee space is maintained, and this typically results in less buildup of burr comb in the space between the top of the frames and the bottom of the inner cover. Because I prefer to use an escape board rather than a bee escape when removing the harvestable honey from the hive, a single oval-shaped opening in the inner cover is unnecessary for me. As a result, I like to drill two 1- to 1¼-inch holes in the inner cover. Not only does this provide increased ventilation through the cover, but it allows me the option of being able to feed a hive two buckets of honey at once on those occasions when feeding may be necessary.

Homemade hive equipment can be modified to more closely fit in with your particular beekeeping management style. This inner cover can accommodate two feed buckets at once. Photo by Hilde Whalley.

Feeding

All living organisms, including honey bees, need high-quality nourishment to reach their maximum potential. As with all living things, insects that are well fed are less susceptible to disease than malnourished insects. Pure honey and pollen provide the bee with calories for energy and the protein and minerals required for optimum health. If the bees do not have access to sufficient quantities of honey of their own making, organic honey standards dictate that only certified organic honey and/or sugar be used as supplemental feed. Honey is the preferred nourishment for feeding bees, with sealed combs of honey taking precedence. Extracted honey from a disease-free source comes in a close second, followed by boiled honey syrup from a questionably clean source. When in doubt, it is best to play it safe and mix the honey with a little water and boil it for about two hours in order to kill any disease spores that might be present. If honey is unavailable for feeding bees, sugar syrup is the next best thing. Only white sugar should be used to make sugar syrup to eliminate the indigestible parts of the feed. Raw or brown sugar should never be used unless absolutely necessary, and then only in the spring when the bees can make cleansing flights in order to void the indigestible waste and are unlikely to store the feed for winter use.

Numerous contraptions are available to feed syrup to hives. They are all variations on entrance feeders, top hive feeders, division board feeders, and feeder pails. As their name implies, entrance feeders are placed at the entrance of the hive. Entrance feeders are easy to use, but they require the workers to travel far from the brood nest, and therefore they lose their effectiveness when the temperature drops into the fifties or below, at which time worker mobility is reduced. Entrance feeders allow for easy feed-level checks and refilling compared to other feeder designs, which is a good thing, given that this style of feeder doesn't usually hold a very large amount of syrup.

A diversity of pollen and nectar sources helps ensure that the hive will be able to adequately provide for its nutritional needs. Access to many different types of plants will become even more important as Earth's climate changes accelerate.

Top hive feeders are shaped like a super and placed above the brood nest underneath the inner cover. Although the bees still have to travel a significant distance to reach the food, heat rising up from the brood area facilitates collection activities when temperatures approach levels where an entrance feeder would not be adequate. Top hive feeders have the potential to hold the greatest amount of syrup of all the feeder types, reducing return visits by the beekeeper to refill the feeder.

The division board feeder replaces a frame, or two, in or next to the brood nest. This feeder offers the benefit of locating feed right by the brood area so bees don't have to travel far to reach it. Small pieces of wood floating in the syrup provide a life-saving platform that will allow the overeager bee to drag herself out of the sea of sugar. The biggest drawback to using a division board feeder is the extra work involved with checking the feed level and refilling the feeder. It also reduces the amount of room within the hive that can be used for food storage and egg laying, because normal frames of comb are removed to make room for the feeding apparatus.

Pails, cans, or jars that have a lid perforated with small holes can be filled and placed upside down over the hole in the inner cover to allow the bees to suck down the syrup. An empty hive body placed on the hive with the outer cover on top will protect the feeder from the elements and prevent robbing. Feeder pails typically come in gallon and half-gallon sizes. This approach offers ease of access for both the beekeeper and the bees, because the position of the hole in the inner cover is typically right above the brood nest. The biggest problem with feeder pails is the potential for sugar syrup to drip down and drench the bees at a time when cool weather prevails. Turning the pail upside down and letting the liquid run out until a vacuum is formed, thus stopping the syrup from flowing, before placing it on the hive will help avoid soaking the bees.

Some folks like to fill an empty frame of drawn comb with sugar syrup and place it in the hive. This is best accomplished in the bee yard to reduce spillage during transportation. A paintbrush dragged across the face of the comb will help to break the surface tension and allow the syrup to collect in the cells.

Ensuring that hives are full of honey as they head into the winter months is one of the beekeeper's primary responsibilities.

Plain dry table sugar can also be used, but only as an emergency measure, such as during extremely cold weather, when it is not desirable to open the hive to install or refill a feeder. When sugar is sprinkled around the hole in the inner cover, the bees will be able to use moisture to dissolve the crystals and consume the sugar. The use of sugar candy (fondant) for emergency feeding has also reportedly met with some success. Such candy is typically made by mixing syrup made of sugar and water with a small amount of tartaric acid, which inverts the sugar. This inverted sugar is then combined with powdered sugar until it is transformed into a stiff dough.

Leaving a bulk container of bee feed sitting open in the bee yard with a cover over it to keep out the rain is the simplest method of feeding bees, but it is also the least efficient. Straw, pine needles, or bits of wood must be added to the container to provide a life raft for the feeding bees. Unfortunately, the strongest hives will typically benefit the most from such general feedings, while the weaker hives that need the food most are least able to fatten up on such an offering. This method can also end up providing food for a host of other insects in the area along with the bees, and it also encourages robbing behavior among the honey bee colonies.

Bees need to be fed only when starvation is imminent. This can occur when, due to weather conditions or some other factor, the colony has not stored sufficient honey to last the bees through the winter, or because their keeper has taken more than his or her share of the harvest. Colonies that need to be fed in the spring are typically a sign that the beekeeper has harvested too much honey and did not leave enough on the hives in the fall, that the late-fall honey flow that was anticipated to fill out the empty space within the hive did not materialize and adequate supplemental feeding did not take place, or both. Another instance when a hive typically requires feeding is when a nuc or swarm is starting out with nothing but foundation or empty combs.

Whenever the feeding of hives is taking place, no honey supers that will be harvested should be on the hive. This is simply a matter of integrity. Pure, natural honey is defined as plant nectar that has been collected and transformed by honey bees. The beekeeper who offers "honey" produced from sugar syrup, or any other source, has compromised the integrity of the final product, as well as his or her moral character. It is questionable whether the apiculturist who removes the real honey the bees need for winter and replaces it with "honey" from an inferior source such as processed white sugar, corn syrup, and so forth is deserving of the name "beekeeper." The term "bee exploiter" may be more fitting.

The feeding of pollen substitutes falls into the same category. Bees consume pollen for its high protein and mineral content, and the colony requires an abundant supply when raising brood. The supply of pollen is most critical in late winter/early spring in the Northeast, when the hive's pollen stores are typically running low and the first pollen-producing plants—the soft maples, various willows, and wildflowers—have yet to bloom. As a wild-grown, nutrient-dense food source, pollen is difficult to replicate, though many have tried, typically by using soy powder as a substitute. Patties of soy flour are fed to hives early in spring to jump-start brood rearing in an effort to ensure a maximum worker population that will take full advantage of the first major honey flow of the season. However, an evaluation of the nutrient and protein differences between pollen and soy reveals that imitation pollen patties are a poor replacement for the real thing. Whether one is replacing honey with sugar syrup, or pollen with soy flour, the overall health and immunity of the colony will be weakened. This will occur even if no noticeable effects are immediately evident. Because of this, I believe that nonorganic and commercial operations that utilize such feeding techniques will tend to experience more problems among their hives than those that don't, especially if sugar and soy supplements are used over prolonged periods. Common sense dictates that, like all creatures, honey bees thrive on wholesome natural foods, and they suffer when forced to consume the equivalent of junk food.

Reversing

Reversing a hive is a technique used primarily to reduce, but not eliminate, a colony's swarming impulse. Preventing an early swarm helps to ensure that the bees will have time to stow away enough honey to last the winter before they divide themselves up when swarming later in the season and thus reduce their nectar-gathering abilities. This can be crucial during poor honey years when the total amount of honey produced by the hive is low. Reversing also provides an opportunity for the beekeeper to conduct a thorough inspection of each colony, so that potential issues may be resolved before they become acute.

Just like it sounds, the process of reversing a colony simply involves restacking the boxes that make up the hive so that the bottom hive body ends up on the top, and vice versa. The intention is to end up with the brood and honey positioned as close to the bottom of the hive as possible, with extra space above. When using a system where the main hive consists of two shallows with a hive body sandwiched between them, the positions of the shallows are reversed. If the hive proper consists of two deep hive bodies, then the position of each deep is switched. Given that the process of reversing a colony involves taking the hive apart, this is a good time to clean the equipment up a bit. The following is one example of how to go about reversing, cleaning, and inspecting a hive in early spring.

First of all, when you walk up to a hive, it is best to approach from either the side or the rear of the colony. It's not a good idea to position yourself directly in front of a hive, because the foragers that are coming and going from the front entrance may interpret your actions as trying to block their path and may feel threatened enough to defend

In northern climates, reversing the order of the hive's chambers in spring, before they have become overcrowded, can delay the bees' swarming instinct, helping to ensure that the colony will have the workforce available to store enough honey to see them safely through the winter.

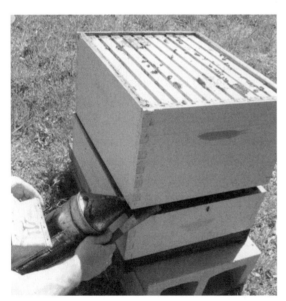

A well-built smoker can aid the beekeeper by acting as a wedge that prevents supers from collapsing together once separated and crushing bees. Photo by Hilde Whalley.

their turf with their stingers. After blowing some smoke into all the entrances of the hive, lift the outer cover and send a puff or two of smoke under the cover before removing it. Place the outer cover on the ground a couple feet behind the hive with the bottom side up. After applying a puff of smoke to the bees gathered around the opening of the inner cover, use your hive tool to pry up a corner of the top super or hive body. As your hive tool lifts up the corner of the super, send some smoke into the crack between the supers and continue to pry up until you can stick the entire nose of your smoker into the space between the boxes. This will allow you to remove the hive tool without the two boxes coming back together and possibly crushing some bees in the process. With the nose of the smoker inserted between the supers, squeeze the bellows a couple times while you reach over with your free hand to the handhold on the raised end of the upper chamber and lift it up. If the back end of the top box is too close to the edge of the box underneath it, you may have to slide the hive body being lifted forward a bit to keep it from slipping off the back end of the hive. After allowing some smoke to drift over the exposed bees, the top super with the inner cover still attached can now be lifted off the hive and placed on the telescoping edges of the outer cover that is on the ground, so that there is little chance that the bees on the exposed underside will be crushed. If more than two boxes make up the hive, repeat the procedure as outlined for the top hive body and place the second hive body off to the side, standing on end.

Whenever you handle supers or hive bodies containing bees, it is best to keep them as level as possible. When it is not possible or convenient to keep a box full of bees in a horizontal position, it is important to always tilt the hive section so that the short end faces down toward the ground, thus preventing the

50

frames from collapsing upon one another. If a nuc or hive body is carried or set down with one of the long sides facing the ground, gravity will force the frames together and in the process kill or injure the bees that are unfortunate enough to get pressed between the leaves of honey, like flowers in a book. If the queen is one of the victims, the hive may well suffer a lethal blow at the beekeeper's hand.

Once the hive body sitting on the bottom board has been removed, scrape the bottom board clean of dirt and debris before returning it to its place as the floor of the colony. Now you are ready to rebuild the hive by replacing the hive bodies in reverse order as compared to their previous position. Unless the colony is exceptionally strong, or you're late in getting around to reversing them, the former bottom portion of the hive will likely contain few bees, little honey, and no brood. Next to the bottom board is precisely where you *don't* want additional space to be within the hive. Because the bee's natural inclination is to move upward, an empty hive body or super should be located on top of the hive, and the boxes filled with the most honey, pollen, bees, and brood should be positioned as low in the hive as possible. This way, the underutilized space within the hive is kept to a minimum, and your equipment is used most efficiently, because the lion's share of the honey gathered will be stored above the brood nest in preparation for winter. Sometimes you will find that a colony that went into winter with three hive bodies will have all their bees, brood, and honey confined to the uppermost story of the hive in the springtime. In this case the bottom hive body can be taken away and the middle section placed on top of the top section to provide the needed room for growth.

Positioning the free space within the hive on top simplifies the process of inspecting the colony in the future, since all it takes is a peek under the inner cover to tell whether additional room is needed. Once

Allow a super to slip off the hive during an inspection, and the fun really begins.

The bees appreciate it when the beekeeper makes use of an inverted outer cover to prevent grass, dirt, and debris from getting all over the hive's boxes when separated.
Photo by Hilde Whalley.

a colony has filled all but the outermost frames in the top super or hive body, an empty super should be placed on top to provide additional space. Back in the old days, beekeepers placed empty supers just above the brood nest and beneath the honey supers that the hive had already filled. This is because the bees will tend to draw out foundation and fill empty combs located in the middle of the hive a little faster than when they are placed at the very top of the colony. However, the disruption to the colony, and the labor involved with checking the supers, and then removing and replacing the full honey supers every time an empty super needed to be added, eventually caused this approach to be discontinued in favor of supering from the top. The one exception is during the production of comb honey, when new supers of extra-thin foundation are added below the full supers to reduce the incidence of "travel stain" that occurs when bees constantly walk across the filled and sealed combs to

Allowing the frames of a hive body to collapse together is a good way to squash a queen.

get to the new foundation above. This occurs in much the same way that dirt builds up on carpeting in high-traffic areas of your home. By positioning the new supers below the previously filled ones, the appearance of the final comb honey harvest is greatly improved.

Incidentally, in general, queens prefer to lay their eggs in the larger spans of comb found in the deep hive bodies rather than in the smaller shallow or medium-sized supers. However, this preference will not stop the queen from laying in shallow supers, especially if the brood nest becomes congested and additional space for egg laying becomes difficult to find. Although it is inconvenient to find brood in your honey supers, it is a healthy thing to allow the queen all the room she requires to lay her eggs. A gentle way of encouraging the mother queen to keep her brood confined to the lower part of the hive, without forcing her to with the use of a queen excluder, is to start off the honey supering process with shallow supers. Because the bees will be more likely to fill a shallow super with nectar, rather than eggs, two or three shallows should be built up upon the colony before any deep honey supers are used. The relatively large span of sealed honey provided by several filled and capped shallow honey supers acts as a natural queen excluder and discourages the queen from crossing all that honey to lay eggs in a distant part of the hive.

Getting back to our hive reversal process: Prior to placing the top hive body, which is chock-full of bees and honey and still has the inner cover attached to it, on the bottom board, the bottom of the frames of the hive body should be scraped clean of all burr comb. Care must be taken to scrape off burr comb and brace comb clear down to the wood, since any wax remaining will act as foundation and provide the bees with guidance and motivation for rebuilding in the same location. If you find your hive tool is gouging and peeling off layers of wood while scraping your equipment clean, your tool may be too sharp. Dragging the sharply honed end of the hive tool along a rock or brick a dozen times or so will dull the edge enough so that, when held at the proper angle, it will remove wax and debris without stripping off the wood underneath like a planing tool. By taking the time to thoroughly clean burr comb off the frames and other areas of the hive early in the season, a lot of trouble and inconvenience can be avoided later on. As with the process of cleaning up dead hives, a significant amount of beeswax can be collected for rendering by scraping the tops and bottoms of the frames into a box before setting the hive bodies and supers back into place. This also helps to prevent the crushing of many bees when repositioning the hive bodies on their stand. You would be surprised at how much beeswax can be collected from a hive during such a process.

Wrapping Up the Season

A major concern with keeping bees in northern locations is winter survival, but I think far too much attention is usually given to concerns about the cold temperatures that the hive will have to endure. As long as the hive is strong and healthy with plenty of adult bees and brood, has an abundance of sealed honey stored in the combs, and is able to stay dry, it should have little trouble withstanding the frosty breath of Old Man Winter. Rather than trying to keep the entire interior of the hive warm, the honey bee cluster serves to focus and conserve the colony's energy by keeping warm the minimum space necessary for survival. As such, I find most efforts to insulate and wrap hives to be a waste of time. If I kept my bees in a climate colder than USDA plant hardiness zone 4, where it typically gets down to about -25°F (-32°C) for a maximum of a week or two each winter, then I might view the practice of wrapping or packing hives with insulation differently. As it is, I have never wrapped or packed my hives for winter, and yet, even since the arrival of varroa mites, I can achieve good overwintering results by simply focusing on the three key ingredients of winter survival: strong, healthy colonies; plenty of honey stores; and dry bees.

Because we have already covered ways to ensure a strong, populous colony with plenty of honey stores for winter, we will focus here on the issue of moisture within the hive. A secure, waterproof outer cover and hive bodies that are solid and tight-fitting are the hive's primary defenses against moisture. Many beekeepers will use a rock or brick to weigh down the outer cover so that strong winter winds will not pry the cover loose, flinging it aside and allowing precipitation to enter. This usually works fine unless the apiary is located in an area where farm animals like cows or horses may use the hive as a scratching post and inadvertently push a colony over while rubbing up against it. Wet areas that experience severe frost-heave activity can potentially cause our little bee condos to topple over as well. During major storms, floods or powerful winds can pick up a hive and transport it a considerable distance. To help defend against these types of occurrences, or when you have so many hives that the number of rocks or bricks needed to weigh down the covers is unwieldy, outer covers may be tied down securely with thin rope or twine (see page 56). At a minimum, the twine should wrap around the two uppermost hive bodies and hold them together, to increase the chances of colony survival in the event that the hive becomes dislodged from its base during the winter. If the twine or string used to tie down the covers in the fall is made from a natural material such as jute or hemp that is not treated with chemicals to prevent rot or repel rodents, it can be saved after being removed in the spring and reused as smoker fuel.

> It is questionable whether the apiculturist who removes the real honey the bees need for winter and replaces it with "honey" from an inferior source such as processed white sugar, corn syrup, and so forth is deserving of the name "beekeeper." The term "bee exploiter" may be more fitting.

Another source of moisture that may find its way into the hive during the winter comes in the form of water that runs in the bottom entrance and collects on the bottom board. Placing a piece of wood or some other material under the rear of the hive will tilt the colony forward so that water from rain or melting snow will not run into the hive. This is especially important for colonies with bottom boards that extend out in front of the hive to provide the bees with a landing area, because these extensions can catch a significant amount of snow and rainfall. Colonies outfitted with screened bottoms made of wire mesh or hardware cloth open to the ground don't need to be concerned with this possibility, since any water that runs into the hive will drain through the screen rather than collecting on the bottom board.

A large amount of moisture is created within the hive during wintry weather as a by-product of honey bee respiration. The importance of adequately ventilating the hive to remove this vapor during cold weather cannot be overstated. This is why the reduction in ventilation created by most hive wraps, blankets, or packing materials often causes more harm to the hive than any insulating benefits they provide. Packing insulation can also serve to lock in the cold and prevent the colony from responding to temporary warm periods that would normally allow the cluster to shift its position within the hive and keep in contact with crucial honey stores. The colony's ability to take advantage of short winter thaws is also important because they provide the bees with the opportunity to go on cleansing flights and relieve themselves of the indigestible products of feeding activity. By the same token, the well-packed hive will take longer to warm up in the spring, delaying the ability of foragers to take advantage of the earliest pollen sources that are so crucial to the colony's spring buildup. (For these reasons, it is also preferable to place apiaries where hives will receive direct sunlight during the day.)

To help guarantee adequate airflow during the winter, I do not recommend reducing the opening in the bottom entrance of the hive. This advice will go against much of what you are likely to read in other books, and it certainly runs counter to how we maintain our own living spaces during the winter. In my opinion, though, entrance reducers should be used only during the spring and early summer when a colony, captured swarm, or nuc is small and weak and needs to be protected from robbers. If a weak colony is discovered in the

Tying Down Hives for the Winter

Wrap the cord around the hive so that it catches on the second or third box down from the top. Cross the loose end under and then back over the opposite piece about four to five inches from the end (A). Now, pass the working end down behind the standing part (B) and feed it through the loop created by it and the opposite piece. This ties the loose end around itself in a figure-eight pattern as shown (C). Tightening this knot will create a slipknot that will allow you to pull the twine tight, like a lasso around the hive. Run the loose end over the top of the hive and down the other side, and cut the string so it hangs down a few inches below the horizontal strand. Pass the cut end over and then under the horizontal cord that is strung across the opposite side of the colony (D). Push down on the outer cover while simultaneously pulling the twine tight. Then pull the free end forward and place a thumb or finger on the junction of the cords to prevent them from loosening up. With your free hand, thread the working end across itself and up and under the horizontal piece of twine opposing it (E). Loop the cord and pull it through the hole created by the horizontal strand, the vertical strand, and the working end so that it tightens on itself (F). The cover of your hive is now securely tied down with a knot that quickly and easily comes apart by simply pulling on the loose end like a shoelace.

A recommended way of tying down a hive cover for the winter. Remember, dry bees are happy bees.

it is better to combine it with another hive rather than to try to carry it through the winter by itself, because such efforts typically end in failure. A three-quarter-inch hole drilled into each hive body, below or to the side of the handhold, will provide an upper entrance that can greatly improve ventilation. Should the bottom entrance become clogged with snow, ice, or dead bees, an upper entrance becomes especially important, not just for ventilating, but also to allow egress for the colony. The use of screened bottom boards open to the ground also improves hive ventilation. Adequate ventilation that will help the bees to remove excess moisture from the hive ranks right up there—along with plenty of honey, a large population of healthy bees, a secure hive cover, protection from the wind, and exposure to direct sunlight—as one of the most important wintering requirements for a hive located in a northern climate.

Aside from tying down the outer covers on hives, the only other winter preparation I make is to insert a one-inch-thick sheet of foam insulation between the inner and outer covers. An empty grain bag with a piece of three-inch fiberglass insulation inside of it that has been cut to the size of an inner cover also works well. I have even heard of folks using other materials such as loose straw or dried leaves as insulation. Such material placed between the outer and inner covers acts like a hat, keeping much of the heat created within the hive from leaving as it rises from the winter cluster. More importantly, it helps prevent the moisture riding on the breath of the colony from collecting on the underside of the inner cover, freezing during a cold spell, and then thawing out later on, only to drip back down on the bees. The bee cluster can handle being wet when it is warm, and it can deal with very cold temperatures as long as the cluster can stay dry and maintain its access to honey. However, like with people, if honey bees become wet and are exposed to cold temperatures, it is a recipe for disaster.

Unpacking the Hive in Spring

The first warm, sunny day in early spring when the temperature rises above 50°F (10°C) is a great time to pay a visit to your apiary. This first visit of the year involves primarily a quick check of the hives, simply to make sure they are still alive. A trip to see the bees can provide great peace of mind as one plans for the upcoming season.

When it comes to providing increased hive ventilation for the winter and an emergency exit for the bees, an upper entrance is tough to beat.

If it is early in the day and the hive has not had time to warm up, no bees may be visible around the entrance or the opening in the inner cover. In such instances, a sharp knock on the side of the hive should produce a reply that will tell you if somebody is home. If after knocking you don't hear an answer, a peek underneath the inner cover is usually enough to confirm whether the hive is still occupied. This is also a good time to make sure that the colony has at least several full frames of sealed honey to tide the bees over until the first honey of the new season begins to flow. The "heft test" is an expeditious way to judge whether the bees will need to be fed at this time of year. Although colonies will obviously be lighter in the spring compared to the fall, they should still have considerable weight to them. By getting a good feel for the weight of hives within which you have confirmed that there are still at least four to five capped frames of honey, you can learn to accurately estimate a hive's stores with a quick lift.

One of the most sorrowful parts of beekeeping is the discovery that a colony has died sometime during the winter. As with all dead hives, the first thing to do is conduct an autopsy (see page 61). Knowing why a colony has expired can help prevent the same fate from afflicting other hives, especially if the brood nest shows signs of American foulbrood (AFB). If the equipment is free of disease, however, any supers or frames filled with honey that are found can be used to feed other hives that are low on food, thus saving you the work of cooking up a batch of sugar syrup.

The discovery of a dead hive also provides an opportunity to take care of the important housekeeping detail of cleaning the equipment. This cleaning will prevent the frames from getting so glued up that they become next to impossible to remove. Take out every frame in each of the hive bodies and supers that made up the former colony, scraping away all of the burr comb and propolis attached to the sides and frame rests of each empty box. Then clean the excess wax and propolis off each frame before placing it back in position. The most efficient way to remove frames from a hive body in which the frames have been securely propolized into place is to drop the empty hive body upside down on an upturned outer cover so that the box comes to rest on the telescoping edges of the cover, and the frames are able to drop down into the space below.

While cleaning up a dead colony, it's a good idea to collect the scrapings from the equipment as you work. With the price of wax and propolis significantly higher than the price of honey these days, the value of these precious substances should not be over-looked. An easy way to gather these items is to scrape the equipment above a collection box that will catch the debris as it falls. A simple version can be made by attaching a piece of window screen to the bottom of a hive body. The screen will allow honey and water to drain, while retaining the propolis and wax pieces within the box.

Heavily propolized frames can be difficult to remove when cleaning out a dead hive. By turning the hive body upside-down and dropping it on an inverted outer cover, gravity will do the work for you.

I like to wait until daytime temperatures have reached 50°F (10°C) or more, on a fairly consistent basis, before I untie the colonies and remove the insulation underneath the outer cover. If a colony is coming out of the winter in a weak condition with only a handful of bees, leaving the insulation in place a little longer will help the colony weather the cool nights until it's time to reverse the hives. The transition from winter to spring is the time of year when honey bees in the North Country are at their most vulnerable. Hives coming out of winter have been weakened from a significant loss in numbers as individual bees have died of old age or other causes after being cooped up inside for most of the past six months. The temperature swings that occur during this season can also wreak havoc on the bees as they invest precious energy and limited resources in expanding the brood nest during warm spells, only to have the cold return with a vengeance. Due to their decreased population, the bees are often unable to adequately cover the expanding nest area and keep it warm during cold snaps, resulting in the brood getting chilled and dying, a condition known as chill brood.

This is also the time of year that can be lethal should the bees lose contact with their stored honey. On too many occasions, I have inspected the brood nest of a dead colony that still had plenty of honey stored within its combs, but the honey was located several inches away from where the bees were clustered. The telltale sign that the bees starved from want of nourishment is that many of them can be found with their tails sticking out of the cells as they scrape the back of the cells that form the comb with their tongues, desperately looking for a drop of golden ambrosia. This is why the first visit of the year to the apiary is so important. Should you find, upon initial inspection, a colony that appears to have eaten itself into a corner, and that the brood area is surrounded by empty comb, move a frame full of sealed honey over so it sits next to the cluster. By making sure that the bees are free from disease and have plenty of honey within reach, many unnecessary deaths can be avoided during the critical transition from winter to spring.

Notes

1. M. Beekman and F. L. W. Ratnieks, "Long-range Foraging by the Honey-bee, *Apis Mellifera* L.," British Ecological Society, Department of Animal and Plant Sciences, University of Sheffield, *Functional Ecology* 14 (2000): 490–96.

Bee Hive Autopsy Results

Characteristics/Symptoms	Probable Cause
Few or no adult bees left in hive either on combs or on the bottom board. (see glossary)	Colony Collapse Disorder
Pupal mass under cappings is brownish in color and has a ropey or elastic viscosity. Sunken brood cell cappings that are dark brown or black in color and have a greasy appearance. Some cappings may also contain small pin holes. Comb have a distinct foul smell.	American Foulbrood (*Paenibacillus* larvae)
Small pin holes in brood cell cappings. Numerous dead bees with deformed wings and/or short abdomens. Numerous dead varroa mites found on bees, in sealed brood cells or on the bottom board.	Varroa mites (*Varroa destructor*) and associated parasitic mite syndrome (PMS)
Remains of dead cluster contain bees that are positioned head-first in cells. Any honey left in the hive is usually located two or more inches away from the remains of the cluster.	Starvation
Remains of numerous drone brood cells sometimes scattered within worker brood on the same comb.	Old or Failed Queen/Drone Layer
Combs, brood, or dead bees covered with mold or mildew.	Indicates that the hive died a while ago or was too weak to maintain combs
No honey left in hive. Wax cappings that covered areas where honey was stored have been ripped open—jagged capping pieces litter the bottom board.	Hive died out or was too weak to defend honey stores from robbing by other bees, wasps, and/or hornets
Significant brown spotting or large patches of brown staining on frames, comb, or the in front of the hive.	Nosema disease or dysentery
Numerous dead bees lying in front of the hive maybe combined with brown spotting on inside or outside of hive entrance. Bees that have disconnected their two pair of wings and rotated them into an orientation that resembles the letter K.	Tracheal Mites (*Acarpis woodi*)
Buildup of webbing on combs containing small black pieces of debris. Remains of old cocoons and rounded elongated indentations in wooden ware. Damaged/disintegrated combs. Grey moths either dead or alive.	Greater or Lesser wax moths (*Galleria mellonella, Achroia grisella*) moved in once the hive became too weak to defend itself or died out
Small hard larval remains that are white, gray, or black within the brood comb or on the bottom board.	Chalk Brood (*Ascosphaera apis*)
Combs are riddled with holes (but no webbing is evident). Inside of hive is covered with slime and any honey left in the hive is fermented and runny. Some SHB larva may still be present.	Small Hive Beetle (*Aethina tumida*) moved in once the hive died out or became too weak to protect all areas of the comb
Sudden collapse of hive. Numerous dead bees lying around in front of the hive with their tongues sticking out.	Pesticide/Chemical Poisoning

CHAPTER 4

GENETICS AND BREEDING

Make boot upon the Summer's velvet buds,
Which pillage they with merry march bring home
To the tent royal of their emperor:
Who, busied in his majesty, surveys
The singing masons, building roofs of gold.

—William Shakespeare

As with all disease, when working with bees, "an ounce of prevention is worth a pound of cure." This is more than just a trite saying. Organic farmers know that the soil is the place to begin in their efforts to produce a successful crop and prevent crop failure from disease and plant malnutrition. Healthy soil is required to grow strong healthy plants. The more robust the plants, the higher the quality of the crop harvested from those plants. The beekeeper, however, is not tied to a single plot of land and the soil it contains but instead must rely on the genetic "ground" within the hive—in other words, the queen, or mother bee. I have come to firmly believe in the wisdom shared with me by Charles Mraz that the more genetic diversity that exists in your hives, the better. Charlie believed, as I do, that raising "mongrels" rather than "thoroughbreds" gives a greater depth of genetic diversity that will give the bees an edge when dealing with unexpected challenges. As a result, over the years I have purchased and introduced into my apiaries several strains of honey bees with mite- and disease-resistant tendencies. The apiculturist who maintains pure strains of bees in all of his or her hives can be compared to the farmer who grows only a single strain of one type of crop. This type of farming is known as a monoculture, and unless you are a researcher or a queen breeder dedicated to maintaining a specific breed of bee, keeping a pure strain of honey bee may leave your hives unnecessarily vulnerable.

The Irish Potato Famine is a poignant example of the folly of relying on a monoculture. During the course of the two centuries leading up to 1845, the poor families of Ireland had become increasingly dependent on the potato to feed themselves. Potatoes provided good nutrition, were easy to grow, and required a minimum of knowledge, labor, and technology to produce, and they kept well when properly stored. All this made the spud a favorite of the lower classes in Ireland. Over time, Irish farmers came to rely increasingly on the most productive potato varieties available. It has been estimated that, in 1845, only two potato varieties accounted for the majority of the spuds planted in Ireland that year. Unfortunately, these varieties were highly susceptible to a blight that left acre upon acre of Irish farmland covered with black rot between 1845 and 1850. Over a million Irish people are said to have perished during the Great Potato Famine of the 1840s.

The apiculturist can learn much from this tragedy. By keeping honey bees of varied genetic stock in their bee yards, beekeepers can help to ensure that at least some of their bees are likely to contain traits that will help them survive outside stresses that may come along to threaten the viability of the colony.

Although some queen breeders make use of isolated breeding yards, purebred honey bees are typically obtained through the practice of artificial insemination. Unfortunately, artificially inseminated queens seem to exhibit more problems than naturally mated queens. It has been my experience that, shortly after an artificially inseminated queen has been introduced into a queenless colony, a hive will often raise a new queen through the supersedure process to replace the artificially inseminated one. Thus, to help ensure the longevity of the genetic characteristics contained within an artificially inseminated queen that has been transferred into a bee population, it is wise to raise daughter queens from the eggs laid by these mother bees as early as possible. Such actions are not as urgent when dealing with naturally mated queens, who tend to lay more eggs on a daily basis and live longer than queens mated in a laboratory. Although at least some of the genetic traits of an artificially inseminated queen are likely to continue on in the hive that successfully replaces her through supersedure, such a loss can certainly be demoralizing to the beekeeper. This is especially true in the case of "breeder" queens, which can cost $250 or more. In addition, supersedure may be accompanied by a decrease in egg laying or an interruption in the brood cycle. This can slow the population growth of the hive and may adversely impact the colony's ability to store enough honey and provide for itself during the winter months. Common sense—which is not so common anymore—tells us that bees know how to inseminate and raise queen bees better than humans do. To me, it makes eco*nomical* as well as *ecological* sense to let the bees raise their own mothers.

Swarm cells that house a queen tend to be built along the bottom edge of the comb, into which the queen will lay her eggs. Supersedure queen cells tend to be built around an egg that the queen has laid within worker comb. Photo by Hilde Whalley.

This is not to imply that artificially inseminated queens cannot be as good as, or even better than, open-mated queens. It's just that with artificial insemination there is a human factor that enters the equation, a factor that does not exist with queens that mate naturally. This additional factor, when added to the long list of small miracles that must take place for the successful enthronement of a queen mother, brings with it additional complications and the possibility of failure. The potential is created for something to go wrong, whether due to human error, poor-quality work during the insemination process, an oversight, or some unknown or poorly understood factor that silently makes its presence felt when the honey bee's natural mating process is bypassed.

As a great appreciator of honey bees, I like to consider myself a steward of the hives in my care, and I make an effort not only to keep hives alive and thriving, but also to avoid crushing or killing individual bees when working with them. By the same token, I do not believe in purposely killing and replacing aged or poorly performing queens. Although a particular queen may not possess certain characteristics that are deemed important to us, such as high honey production, she may harbor traits that are extremely beneficial, but not as obvious. Resistance to mites and diseases are two examples. Out of respect for the bees that provide so many benefits to the rest of us that share this wonderful blue-green planet, my resistance to purposely killing a queen mother and replacing her simply because she reaches a certain age, or because she does not express certain traits that I consider worthwhile or valuable, is an example of the stewardship approach to apiculture that I prefer to practice.

In keeping with the philosophy that the more genetic diversity within the apiary, the better, I will wait until a hive goes queenless or the queen becomes a drone layer before introducing a new queen, an unhatched queen cell, or a frame containing unhatched eggs and brood from which the bees will raise a new mother. This gives the weak or poorly performing colony every opportunity either to regain its strength on its own or to replace its leader with a new and vigorous queen. There are many beekeepers who would vehemently disagree with this approach, however, and who argue that requeening a colony that is struggling is not only more profitable in the long run, but the most humane thing you can do for the bees. It seems to me that this position assumes we know better than the bees what is best for the hive when it comes to raising and installing the colony's reproductive organ, and I am not willing to make that assumption.

This interventionist way of thinking reveals itself in the act of clipping the wings of a queen, in an attempt to stop her from flying off with a swarm, or spaying a hive by squashing the one member responsible for its reproductive activity to replace her with a younger, more fertile version. By cultivating the change in attitude that will allow us to treat the honey bee with greater reverence and respect, we nurture the same mind-set that provides hope for the future of human culture. In essence, if we want to create a different world to live in, we have to start by changing our minds. In the universe of the honey bee, the workers will carry out the process of supersedure to replace an old, failing, or injured queen and raise a new queen that will inherit the throne. Should the hive fail to raise a new mother, the beekeeper can respond by simply combining the remaining bees with another hive.

This provides the opportunity to help a hive that is weak and in need of a population boost. This approach can also be used with drone-laying queens, late in the summer, when the chances of raising a new queen and building up a healthy population of bees large enough to make it through the winter are doubtful. A queen becomes a drone layer either by failing to mate or by running out of the male sperm stored within her abdomen following her mating flights. Without sperm to fertilize her eggs, she is able to lay only infertile eggs that will develop into drones.

In essence, we can treat the hive body and supers of the failed queen like empty equipment and use them for supering colonies in need of more room, while the attending bees within the equipment will be accepted by their foster mother and able to join forces with the hive the supers are placed on. Many books will tell you to find the queen within the failing hive and kill her before combining colonies. This can be difficult, if not nearly impossible in the case of a hive that has been taken over by one or more laying workers, because the laying workers are not that much larger than their sisters. Unfortunately, drone-laying hives rarely accept a new queen that is introduced into their hive directly. A queen-right colony (with a healthy, fertile laying queen) will, even when weak, typically be able to accept a hive body containing a drone-layer queen and her attendant workers and drones with little difficulty. It is also possible to encourage the workers in a drone-laying colony to raise a new queen from freshly laid eggs that are inserted into the hive after being taken from a queen-right colony. This procedure tends to work best when carried out early on, when the queen first begins to fail, rather than later, after some time has passed and her drone-rearing ways have become firmly established within the hive. It is almost as if, after some time, the workers left within the colony give up trying to raise a new queen from the fresh eggs laid by their failing mother and are thus unable to capitalize upon the fertile eggs donated by another hive. Compared to killing the failing queen or laying workers and then requeening, allowing the hive to raise its own replacement queen, or at least combining the remaining bees with another hive, is in greater alignment with a basic respect for nature's biological processes and the bees themselves. Such respect has always existed at the heart of the organic movement.

A Case for Locally Raised Queens

No matter what the qualities of a hive are, the beekeeper cannot fully benefit from its specific genetic traits unless the bees can thrive in the climate where the beekeeper has located them. Living in Vermont, about seventy-five miles from the Canadian border, I have found that bees raised in the western and southern United States, where the majority of American queen breeders operate, are not necessarily well suited to the Northeast climate. The same may be said of bees and queens imported from other countries. This seems especially true in the effort to find strains of honey bees that contain a natural resistance to mites and diseases. Just because bees show some mite or disease resistance when bred in warm climates like California or the southern United States, where hives are active on a more or less year-round basis, does not necessarily mean that those same bees are capable of surviving side by side with mites or disease in northern climates. Rather than hibernating like many other insects and animals, the honey bee becomes dormant to conserve resources until spring. A key survival trait for northern bees is their ability to raise a large population of healthy bees in autumn, capable of surviving the winter's period of reduced brood production. Unlike its warm-weather cousin, the Africanized honey bee, the European honey bee has fully developed this mechanism for surviving the cold winter months in northern climes. These late-season bees, born in autumn, have the capacity to live for several months within the relatively inactive environment of the winter cluster. To stay warm, the bees eat the honey the hive gathered during the summer and use these calories to flex their wing muscles and generate body heat. The bees literally snuggle with one another throughout the winter by huddling close and creating an insulating shell with their bodies, to the point where they form a warm ball surrounding the queen. As the bees on the exterior of the cluster cool down, they slowly work their way into the interior to warm themselves, while warmer bees nearer the center move to the outside to take their place. Although the cluster does not heat the unoccupied space within the hive to any great degree, it can regulate the temperature in the middle of the group—where the queen remains—by contracting or expanding the cluster size.

As the cold weather comes on, the cluster will contract, and the queen slows down her egg laying to the point where she eventually may stop laying altogether for one or more periods before resuming her reproductive duties and slowly building the hive's population back up in preparation for the spring bloom. This reduction in egg laying serves two distinct purposes. The first is to prevent the possibility of chilled brood, which can be fatal for the baby bees. Chilled brood occurs when the hive does not have enough bees to cover the developing brood and keep it warm with their body heat. The reduced egg laying also conserves the hive's resources, because it takes significant amounts of honey and pollen to raise baby bees into adults. Reducing the need to consume pollen and honey helps ensure that the colony will have enough provisions to carry it through the winter.

> The more genetic diversity that exists in your hives, the better. . . . Raising "mongrels" rather than "thoroughbreds" gives a greater depth of genetic diversity that will give the bees an edge when dealing with unexpected challenges.

Honey bees born in late autumn have special attributes that make them much more suited to winter survival than their summer-born sisters, who literally wear themselves out by working so much that they may live for only five to six weeks. The fact that honey bees have a hotter sting in the fall than during other seasons indicates that the bees are indeed different at this time of year. Unfortunately, winter survival can be especially challenging for these bees when they have to share their hive with mites. Honey bees are weakened both by mites feeding on them during their development and in adulthood and by associated viruses or diseases the mites may spread through the hive's population.

The best way to build *Apis mellifera* bee stock that is acclimated and well suited to survive the unique conditions of one's area is to raise succeeding generations of them in that same locale. Luckily a lot of specialized knowledge and equipment are not necessary to do this. Good results may be obtained by breeding from one's best hives and working with the honey bees' natural instinct to raise a queen when the hive finds itself without one. Reserving the option to rear queens that have performed well for you in your area by creating new colonies from existing beehives is yet another reason for not automatically requeening hives, as some authorities recommend. In the days before varroa mites, beekeepers would typically split the hives that made the most honey.

Another reason for splitting a colony would have been to reduce the hive's population during overcrowded conditions in an effort to avoid having the bees swarm. With the advent of varroa, however, high honey production ceased being the primary trait beekeepers focus on when deciding which hives to use as breeding stock. Hives that have survived the winter and have a relatively high population of bees in the spring are the colonies that are bred these days, in an effort to increase the number of hives with a genetic disposition for survival in spite of the varroa mite and its associated diseases.

When a hive is divided to create one or more new colonies, each new hive is given a small version, or nucleus, of what is found in a full-sized hive with which to build upon. Thus, the term for these artificial swarms is *nucleus colony,* or *nuc* for short. The beauty of using the existing hives in your apiary to create more colonies is that the potential of importing into your bee population diseases, pests, or unwanted genetic traits, such as those found in the Africanized honey bee, is greatly removed. Some queen breeders are utilizing nuc making in combination with the elimination of mite treatments altogether in an effort to develop varroa-resistant strains of honey bees. It is clear to me that a strong nucleus colony program is integral to the success of this "tough love" approach.

One queen breeder who is working with nucs to develop a mite-resistant bee is Kirk Webster of Champlain Valley Bees and Queens, located in Middlebury and Bridport, Vermont. Kirk's experience has proven that the "live and let die" method is not for the faint of heart. A full-time beekeeper with about a thousand hives at the peak of his season, Kirk has been slowly weaning his bees off varroa treatments, to the point where he doesn't treat his hives for *Varroa destructor* at all anymore. Without his intervention, the hives at Champlain Valley Bees and Queens have experienced winter losses of 60 percent or more in recent years. Kirk's approach is to split and raise queens from his surviving colonies each spring and repeat the cycle over and over again. The hope is that, over enough generations, the bees in his care will self-select for varroa-resistant traits, to the point where a bee that is immune, or at least highly tolerant of mites, will result. Although this is not an attractive option for all beekeepers, this radical strategy certainly has the potential to separate the wannabees from those that are truly resistant to *V. destructor*. However, even if efforts such as Kirk's do not result in true immunity, the benefit of any incremental increases in varroa resistance that may be developed is likely to prove invaluable in the long run.

As with all things, the cold-turkey approach also has its shortcomings. Aside from the obvious emotional and economic stress that may be caused from the loss of a large percentage of hives each year, and the additional labor involved with cleaning out dead colonies each spring, there is no guarantee that this approach will ultimately prove successful. Being as much an art as it is a science, the craft of beekeeping involves a staggering number of variables. Honey bee genetics, atmospheric weather patterns, local flora and fauna, and the type and timing of hive manipulations are just a few of the vast number of influences that can affect the health and survival of a hive.

In this day and age, when varroa mites are so prevalent, one management technique that has increased the likelihood of honey bee winter survivability in hives is to make splits, or nucleus colonies, from the surviving bees. Not only do these splits carry on the genetics of the hives that are already proven to have a level of survivability, but the act of making nucleus colonies, in and of itself, consistently imparts a higher survival rate for the nucs over the winter than for the older "full-grown" colonies. Splitting up a hive to make a nuc not only divides the number of bees in the hive but also reduces the number of mites at the same time, because some are removed from the original hive when the nucleus colony is made. Splitting a hive and then allowing the bees to raise their own queen also interrupt the colony's natural brood cycle, upon which the varroa mites depend. The result is an interruption in the reproductive cycle of the mites as well. By interrupting the brood cycle of the mites through the creation of a nuc that must raise its own queen from an egg, we slow down the exponential growth curve of the varroa population and reduce the likelihood that the number of mites in the hive will build to a level that will cause the colony to collapse later in the season. Indeed, a higher tendency to swarm is one of the characteristics attributable to the mite resistance found in both the Africanized honey bee and certain Russian breeds. Some beekeepers achieve similar results by caging the queen for a while, thus creating a short broodless period. Thus, the ultimate experiment may be to see whether a strong hive can survive for several years without mite treatments, *and* without artificial broodless periods being created within the colony. This may require that today's beekeepers and breeders become more tolerant of the natural swarming instinct within their colonies, a trait that most have long tried to breed out of their bee stocks in an effort to maximize honey production and profits.

The naturally emitted swarm is *A. mellifera*'s way of giving birth to a new colony, and it is one of the most fascinating activities in which honey bees engage. It stimulates the human imagination and stirs the soul to witness a swarm issuing forth from a hive in search of a new home. The queen mother and roughly half the workers within a colony will time their departure so that it takes place shortly before one or more new virgin queens, which have been lovingly raised and cared for, are set to hatch out of their birthing cells and take over leadership of the hive. The confidence the swarm shows in itself and the faith it has in the universe's ability to provide for its future well-being are inspiring. Certainly the mother and her daughters do not make the decision to swarm carelessly, typically timing their move to coincide with conditions when the hive is chockfull of bees, honey, pollen, and brood, and there is plenty of nectar still to be gathered. It makes good

> Being as much an art as it is a science, the craft of beekeeping involves a staggering number of variables. Honey bee genetics, atmospheric weather patterns, local flora and fauna, and the type and timing of hive manipulations are just a few of the vast number of influences that can affect the health and survival of a hive.

sense, after all, to wait until times are plentiful and prosperous before deciding to build a new home and start another family. However, when compared to human parents, the widowed mother bee shows unparalleled generosity of spirit in her actions.

When humans establish a home, become prosperous, and have raised their offspring to adulthood, the tendency is to push the kids out of the nest, so they either learn to fly on their own or hit the ground trying. By leaving behind the hive she has guided to maturation and prosperity, along with her most loyal family members, the queen bee exhibits courage truly worthy of royalty, transferring ownership of her entire estate to the daughters she has left, and in the process ensuring their continued well-being. Flying away from the safety and security of the hive toward the unknown, the swarm leaves behind all the material possessions it can't carry with it. The selfless-ness and generosity exhibited by the honey bees' act of swarming provides a level of inspiration unmatched in the world of insects.

Although the idea of creating artificial swarms and rearing new colonies from your own bee stocks may seem intimidating, it is relatively easy provided you keep in mind the basic biology of *Apis mellifera*. When a hive finds itself without a mother, the workers will go to work and attempt to raise a new queen in short order. The primary requirements to accomplish such a task are a few freshly laid fertile eggs, ample supplies of honey and pollen, enough bees of all age groups to survive the roughly eight-week period before the freshly hatched brood from the newly mated queen emerges from its birthing cells, and enough good weather during the third and fourth weeks for the new queen to make one or more successful mating flights. With thoughtful planning and attention to detail, the beekeeper can often orchestrate the opportunity for the necessities of enough eggs, bees, and good weather to come together successfully.

A well-known example of a strain of bees bred for a specific trait is the Buckfast bee. These bees are genetically resistant to tracheal mites (*Acarapis woodi*) and were developed at England's Buckfast Abbey by the famous beekeeper Brother Adam (1898–1996). In recent years, apicultural researchers have made genetics the focus of much of their efforts to control the damage resulting from varroa mites. Hygienic behavior has become the most studied genetically determined characteristic of honey bees that exhibit varroa resistance. This in turn has resulted in the commercial availability of several bee strains with varying levels of resistance to varroa. These include bees that can trace their genetics to western Russia, near Vladivostok along the Pacific Coast (Primorski stock); the survivor stocks from across the United States that were collected and refined by Dr. John Harbo and Dr. Roger Hoopingarner; and the Minnesota Hygienic stock developed by Dr. Marla Spivak of the University of Minnesota. The Russian breed has been shown to control mites at least partly through increased grooming behavior. Grooming behaviors that fend off the mites include bees biting mites that are on either their own bodies or the bodies of other bees. Should a bee chew off a mite's leg, the break in the mite's exoskeleton will cause the parasite to quickly dehydrate and die.

It is similar hygienic characteristics that make these bees more resistant to other hive diseases as well. The Minnesota hygenic stock and the bees that were bred from various survivor stocks—hives that were able to survive despite the onslaught of varroa—have developed a hygienic mechanism for removing diseases and mite-infested brood from their nest. When applied to mite-parasitized broods this behavior was initially labeled suppressed mite reproduction (SMR). This genetic trait has since been renamed varroa-sensitive hygiene (VSH). VSH bees exhibit the ability to detect varroa-infested brood within the sealed birthing cell and remove it. These bees are also more selective about removing brood infested with mites that are capable of reproducing, and passing over brood that share cells with mites that are unable to reproduce.[1] How the VSH bee is able to differentiate between mites that can reproduce and mites that can't is still the subject of study. However, at this time it seems that the best chance of breeding a honey bee that is truly immune, rather than just being tolerant of or resistant to varroa, lies in combining these types of hygienic behaviors with a strong swarming instinct.

For those who don't have the stomach for the radical measures of the "live and let die" strategy, a more systematic approach has been developed to test for hygienic behavior. Basically, hives are tested to see how rapidly they are able to detect, uncap, and then remove dead or damaged honey bee brood. There are several variations of this test that have been developed over the years. One is to cut out a section of sealed brood, freeze it, and then replace it within the frame in the hive. The frame with the dead brood is then inspected twenty-four to forty-eight hours later, and the number of cells that have been opened and have had the dead brood removed are recorded. The downside to this technique is that it is laborious and very disruptive to the colony. One could even make the argument that any hive that didn't notice something wrong after going through this procedure is *really* deficient. A more refined approach has been developed by Marla Spivak using liquid nitrogen to freeze the sealed brood without removing it from the frame. A hollow cylinder, such as a soup can or PVC pipe, open on both ends, is driven through the brood to the middle of the comb. The rim should be smooth, with all the rough edges filed down to provide a tight seal between the cylinder and the wax midrib of the frame. A good seal is essential for preventing leakage of liquid nitrogen from within the test area. The liquid nitrogen is then poured into the can or cylinder and onto the frame, instantly freezing and killing the brood. Again, the percentage of dead bee brood removed in a twenty-four to forty-eight-hour period is observed. The major shortcomings to this approach are the difficulty in obtaining liquid nitrogen and the dangers associated with its handling.

A simpler technique to conduct the hygienic test uses a pin or needle to fatally injure the developing brood. After the needle is inserted into one hundred cells of sealed brood, the results are evaluated twenty-four to forty-eight hours later. Although still labor intensive, no special or hard-to-find equipment and materials are required with this method.

Hygienic behavior is determined to be present if 95 percent or more of the dead brood is removed within the initial forty-eight-hour period. Limiting the sample area to one hundred cells helps to simplify the calculation of percentages. It also helps if all the brood used for the test is grouped together, so that a clear delineation between the dead brood removed and the live unharmed brood on the same frame is evident. Some researchers believe that a hive should be categorized as hygienic only if the test is performed twice and over 95 percent of the dead or damaged brood is removed within the same time span on both occasions.

Either way, it is important to realize that the proportion of workers required for a hive to express hygienic behavior does not have to be exceedingly high to be obvious and effective. Some research has suggested that colonies with only 15 to 50 percent hygienic bees are just as hygienic when confronted with large amounts of dead or damaged brood as colonies with 100 percent hygienic bees. This may prove important, because queens are believed to be able to mate with as many as twenty or more drones, and only a certain proportion of these drones carrying the hygienic gene are needed to produce a hygienic colony. By the same token, our lack of knowledge as to what degree of hygienic behavior is desirable means that it is possible that it may be overdone. Perhaps, in its extreme form, hygienic behavior can be detrimental to the colony. At this time, current thinking holds that selection for hygienic behavior by breeders will produce a far greater resistance, or even true immunity, to *Varroa destructor,* along with diseases like American foulbrood, sac brood, and chalkbrood.

Whether we take the radical "live and let die" approach or a more scientific one, once a hive with favorable genetic traits has been identified, we can breed the colony in an attempt to pass on and improve upon the hive's favorable characteristics. One way to accomplish this is to raise or graft queen cells and then transfer them into nucleus colonies. This can be a highly efficient method for producing many queens. It also requires specialized equipment and skill. As a beekeeper who is continually trying to reduce the amount of equipment I must deal with, I have forgone this approach for a less efficient but simpler and less labor-intensive way to propagate a colony. The technique I prefer is to simply split a hive in order to create one or more nucleus colonies from the mother hive. This is best done during the primary swarming season, which tends to coincide with local honey flows and will vary among geographic locations. By timing breeding efforts to overlap with the natural swarming season of your area, you can take advantage, as the bees do, of the favorable conditions that typically exist during this time: warm average daytime temperatures; an abundance of blossoms for foraging; and enough time left in the season for the new hive to build plenty of combs and fill them with honey, and to build up a large enough bee population to survive the long dearth of winter. In Vermont, a nucleus colony can be created as late as early July and still have a good chance of successfully surviving the winter. However, late-season swarms, be they natural or artificial, almost always require additional feeding in the fall to prevent starvation before spring.

For those persons who are just starting out with bees, or who don't have enough bees to populate a nucleus colony, bees will have to be purchased. Honey bees are usually sold either as a nucleus colony or by the pound. A pound of bees is typically composed of between three thousand and thirty-five hundred workers, and the standard package of bees available from bee supply companies weighs in at three pounds, for a total of about ten thousand bees. Reputable companies will err on the high side and provide extra numbers to make up for bees that may die during transport, as well as the honey and nectar that is in the stomachs of the bees when they are initially shaken into the package. Bees are safely mailed in special boxes approved by UPS and the United States Postal Service. Please note that a package of bees may not contain a queen and will contain only workers, unless a queen has been ordered at the same time as the package. Be sure to inquire with your supplier if a queen is included. Either way, it is best to prepare your equipment and location before the package arrives in the mail.

When you are ready to transfer the bees from the package into their hive, use a hive tool to pry open the top of the package and expose the feed can. Then remove four or five frames from the nuc box, pry the feed can loose from the package, and place the package without the feed can in the hive. Before removing the can, it is a good idea to knock the bees to the bottom of the package by holding the package a few inches above the ground and then gently rapping the box on the ground. Be sure to hang on to the metal tab attached to the queen cage while removing the feed can. Before installing the bees in their new home, place the queen, still in her cage, within the hive as shown in the photo on page 80. Then replace the hive covers and let the bees crawl out of the package and release the queen. The mostly empty package can be removed the next day (any bees still lingering inside can be shaken out) and the frames returned to the hive body. If the queen has not been freed from her cage, she can be manually released, as the bees will have had plenty of time to acclimate themselves to her during their travel time together in the package.

> Flying away from the safety and security of the hive toward the unknown, the swarm leaves behind all the material possessions it can't carry with it. The selflessness and generosity exhibited by the honey bees' act of swarming provides a level of inspiration unmatched in the world of insects.

Alternatively, the package can be turned upside down and the bees can be shaken out of the open package, making sure most of them fall onto the queen cage suspended between the frames. Unfortunately, this process will tend to get the bees riled up, so be sure to wear a veil while doing this. To help reduce the tendency of the bees to fly during this process, you can turn the package on its side, with the screen side up, and brush sugar syrup onto the screen before opening up the package. Once all the bees have been shaken out of the package, cover the hive and don't disturb the bees for several days in order to give them a chance to free the queen so egg laying can commence. It is a good idea to check the nuc two or three days later to remove the empty queen cage, or release the queen if she is still in the cage, and add the tenth frame that was removed to make room for the cage back into the hive body. If the hive is being created without any frames of sealed honey, additional feeding will be necessary to help the colony get started and prevent starvation in the event that the weather turns cold and rainy and the bees do not get the opportunity to forage for nectar.

Creating the Nucleus Colony

Once you have identified your bees' breeding stock and established when the timing is right, you can get down to the business of creating a nuc. As with all aspects of beekeeping, there are numerous ways to successfully go about this. For beekeepers whose management style is to overwinter hives in two deep hive bodies, the easiest method is to separate the hive bodies and provide both boxes with a bottom, inner cover, outer cover, empty honey super or hive body for anticipated honey storage, and entrance reducer. It is important to make sure there are at least a few unhatched eggs in both of the hive bodies so the worker bees in the box without a queen will have the resources to raise a new mother bee. Since initially we may not know which box contains the queen, it is important that some sealed worker brood, honey, pollen, and plenty of bees are also included in both boxes. After creating the nuc, it is best to leave it alone for thirty days before conducting an inspection for the presence of a queen, because any newly formed unhatched queen cells may be damaged if frame manipulations are attempted prior to the virgin queen emerging from her birthing cell. Unfortunately, there is no guarantee that the newly created nuc will be able to raise a queen or, if it does, that the queen mother will be able to successfully mate and return to her hive without getting eaten by a bird or falling victim to some other tragedy.

The hive containing the original queen will not experience a disruption of its brood-rearing cycle and should contain frames with sealed worker brood. Meanwhile, upon the initial thirty-day inspection, the nuc will not contain sealed brood but should have a newly mated queen that is laying her first eggs. Care must be taken not to confuse a laying worker, or nonfertile queen, with a successfully mated queen who is just learning the ropes. On numerous occasions I have observed that the first eggs laid by a new queen may look similar to those of a drone-laying queen or worker. Eggs flopped over on their sides or not positioned in the apex at the back of the cell, as well as more than one egg in a cell, are all part of the learning curve the new queen may go through as she grows into her role within the hive and figures out the nuances of her plumbing. It is the frequency at which such instances are discovered that helps the beekeeper distinguish between the hive with a mated queen and the one with a drone layer or laying worker. Cells with mislaid eggs will be the norm when dealing with the latter, rather than the exception occurring early in the egg-laying career of the former.

This "splitting" method of creating nucleus colonies requires a minimum of hive manipulation and relieves the beekeeper from the work of having to locate the queen prior to dividing the hive. However, this should be tried only with a hive that has a large population of bees occupying both chambers of the original full-sized colony. This is because the older field bees that leave the nucleus colony to gather nectar and pollen will follow recognizable landmarks and return to the location of the original hive after completing their foraging activities. Therefore, it is important to be sure there will be enough young bees left behind in the nuc to carry on the essential activities of the developing hive once all their older sisters have departed.

One way to prevent the reduction in population caused by the drifting of foraging-age worker bees from the nuc back to the mother colony is to move the nucleus colony to a new location over two miles away from the original hive. This will force the foraging bees to reorient themselves and become familiar with the strange new territory surrounding their hive location, memorizing the landmarks that will lead them back to their newly created home location. However, although this is a great way to improve the chances of a newly created hive's success, it is not always practical or possible to move a nuc to a new location two or more miles away.

Established hives may be moved a short distance within an apiary without the loss of their foraging force by using one of two methods. The first is to move the hive over two miles away for a period of about two to three weeks as described above. Then, once all the older bees of foraging age that would recognize the local landmarks have either died off or reprogrammed themselves, the hive may be moved back into the original apiary and placed in its new location. Alternatively, a hive may be moved a very short distance (one to two feet) away from its original location without the foraging bees becoming so disoriented that they are unable to find their way back home. Regularly moving a hive such short distances and providing a period of a few days between each move can progressively transport a colony to a completely new location within a bee yard without a significant loss of foragers.

Fortunately, leaving the baby colony in the same bee yard as the mother hive, and simply ensuring that there are plenty of young worker bees in the nuc to carry out the hive duties, can still obtain good results. It is important to reduce the entrance down to a single small hole or two with a small piece of wood or some wads of grass to help reduce the potential for robbing should the honey flow end or fail to materialize altogether. During years with a poor honey flow it is possible for a nuc to get totally robbed of all its honey stores due to the bees' weakened and disorganized state, all because the hive entrance was not reduced to allow the guard bees to easily defend the colony.

When you wish to bring specific genetic traits into your bee yard, a queen-right nucleus can be created quickly by introducing a queen instead of waiting a month or so for the bees to raise their own. This method has the ben-

Queen cages can be reused by simply replacing the sugar candy in one end.

efit of gaining an extra four weeks of brood rearing, resulting in the faster buildup of the hive. Although it is possible to introduce a queen directly into a nucleus colony that has just been created, complications can be reduced if the hive is left queenless for at least twenty-four hours before a new queen mother with the desired traits is placed into the fledgling colony.

The best way to introduce a queen into a colony is in a queen cage. The cage slows down the introductory process so that the bees have a chance to become familiar with, and accept, the queen prior to her attempt to integrate into the hive. To introduce a new mother into a hive or nuc, remove the cork from the candy end of the cage. I like to poke a small hole through the candy with a nail or some other similar object, like the reamer/threading tool on a Swiss Army knife. Although it is not necessary, and many beekeepers don't bother to do so, the hole encourages the bees to chew through the candy, thus releasing the queen more quickly.

I like to position the cage between two frames in the nuc, preferably with the screen side facing the bottom of the hive, and tilted so that the candy end is a little lower than the area occupied by the queen. This will make it easier for the bees to make contact with and feed the queen through the screen until the candy plug is eaten away. Unfortunately, the cage will not fit between the frames of a ten-frame hive body without damaging the comb. A frame (preferably one that does not contain brood) can be removed to make room for the cage to fit without crushing any sections of comb. Be sure the cage is wedged in tightly between the frame's top bars because the queen may die if she falls to the bottom of the hive. It is preferable to position the cage between two frames of brood within the colony so that the queen will be located within the cluster, the warmest area of the hive. Once the queen cage is in position, close up the hive and wait three to four days. When you open up the colony again and find the cage empty, look for freshly laid eggs in the combs as an indication that the queen has successfully been released and has been integrated into the hive. If the queen has not yet been freed from her cage, you should release her by removing the screen on the cage and letting her wander out into her new abode.

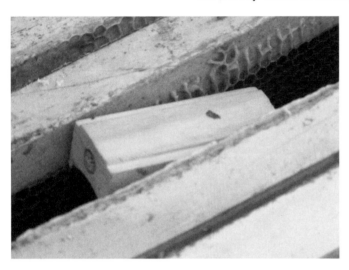

Positioning the queen cage with the screen side facing down and the candy end lower than the corked end takes advantage of the bees' natural tendency to move upward.

Once you have gained some experience and are feeling more confident in your abilities, you can make two or more nucs from a single hive body filled with bees, brood, honey, and pollen. This can be accomplished by removing three or four frames that together contain at least a few unhatched eggs from the parent hive. Properly laid honey bee eggs resemble miniature grains of rice standing up on end and are located way in the back of the cell at the apex or point where the six sides of the cell meet. Be sure that the frames that make up the nuc also contain worker brood in various stages of development, capped honey, pollen, and nurse bees who will care for the young. If the nuc is not going to be moved two or more miles from the mother hive, it is a good idea to shake additional bees from frames of brood into the hive body to ensure that an adequate supply of young nurse-age bees are transferred into the newly created colony. Special care must be taken to ensure that the queen is not among the bees adhering to the frames being transferred or the bees shaken into the nucleus colony. When inspecting each frame for the queen it is best to hold the frame over the hive as it is lifted from its resting place so that, if the queen is on the frame that is removed from the hive and loses her grip on the comb, she will fall back into the hive and not on the ground, where she might accidentally get stepped on.

Bill Mraz of Champlain Valley Apiaries taught me a wonderful system for quickly and efficiently finding the queen, as her swollen abdomen full of eggs causes her to stand out from the rest of the individuals in a hive full of bees. Bill's approach is to position yourself so that the sun is shining over your shoulder and start with the inner cover of the hive. Upon removal of the cover, immediately check the underside for the queen. Although she usually remains within the confines of the brood nest, she sometimes wanders and could be anywhere. In fact, the queen mother is often found up on the inner cover on cold spring mornings in northern climates because this is often the warmest place in the hive at that time of year. If the queen is not on the cover, set it down by leaning it on end against the hive, and turn your attention to the second frame in from the outer edge of the hive. At this time of the year the second frame of comb

It takes about three days before a freshly laid honey bee egg will hatch. Notice the surrounding cells filled with the pollen necessary for feeding the growing young bees.
Photo by Steve Parise.

is usually off to the side of the brood nest, which is typically located in the center of the hive body and is where the queen is usually found. By first removing a frame off to the side of the brood nest, the chances are reduced that the queen will be on that particular frame and get injured as the frame is pulled from the hive, as compared to pulling the first frame directly out of the middle of the brood area. If this is the case, you might logically ask, "Then why not remove the outside frame first, since it is farthest from the brood area?" The reason it is preferable to remove the second frame rather than the first frame is that the center of the comb of the outside frame is often fastened to the side of the hive body with brace comb and will rip loose as the frame is lifted, spilling honey and crushing bees between the broken comb and the side of the hive in the process.

You may have to pry hard to force the frames over in your effort to create enough space to remove that first frame.

To remove the frame that is sitting in the second position, first slice through any propolis (bee glue) that may have built up between the end bars of the first and second frames, and, using your hive tool as a lever, crowd the majority of the frames tightly together against the opposite side so that a little additional space is created between the first and second frames, which makes the removal process easier. The angled end of the hive tool can then be used to pry one end of the second frame away from the frame that is sitting in the third position, raising it up so that you can grab it with your free hand. Once you have a grasp of one end of the frame, pry the opposite end up in the same way and use your index finger to grab this end. The frame can now be lifted from its resting place, slowly and carefully, so that the comb does not rub against the frames on either side. This will require a steady hand and a lifting motion that is straight up, all the while keeping the frame an equal distance from the frames on either side of it. Of course, the judicious use of smoke is recommended throughout this entire procedure to keep the bees in the hive body calm, to chase the bees away from the ends of the frame so that it may be pried up and lifted out of the hive, and to cool the tempers of any bees that get caught between the two frames and receive a rough ride as that first frame is being lifted from the hive.

As soon as the frame that occupied the second position in the hive is lifted from the colony, peek down at the now exposed side of the frame that is sitting in the third position. The queen mother instinctively knows she is safest when she is deep within the dark recesses of the hive, and she has a natural tendency to run away from the light of the sun. If she is on the face of the comb that is exposed, she is relatively easy to spot as she runs down to the bottom edge of the frame and scrambles around to the other side to get out of the sun's glare. If you do not see her on the exposed comb within the hive, look for her on the frame that you have just removed and are holding over the open hive. When looking for the queen on a frame full of busy bees that are bustling to and fro, use a spiral search pattern, starting on the frame's outer edge and working your way inward toward the center of the comb. If the queen is on the frame that has just been removed, she will likely be trying to run from the light and will often be headed for the shaded side of the frame you are holding. By searching the outer edges of the frame first, you are likely to spot her before she has a chance to duck around the corner. If you don't see the queen on the first side of the frame, slowly flip the frame over and repeat the spiral search pattern on the opposite side.

If Her Royal Highness is not found after searching both sides of the frame, set the frame down in an empty hive body, or stand it on end and lean it up against the hive where it will be out of the way, and remove the next frame, the one sitting in the third position. After prying it free from the fourth frame, make use of the space created by the missing second frame to lift out the comb without allowing it to brush close to the other frames, possibly rolling and injuring bees in the process. After taking a peek at the exposed comb on the fourth frame still sitting in the hive, run the spiral search pattern on the frame that was just removed. Keep repeating this search sequence, frame by frame, working your way through the entire brood nest until the queen is found. Remember, although the queen is typically found in the brood area, she could be anywhere within the colony. Therefore, it is important not to place the inspected frames back in the hive, since the sides of the hive body and the bottom board will have to be checked if the queen is not found on the frames.

One last point to keep in mind when looking for the queen is not to use too much smoke. In fact, this is one of the few times in beekeeping that using too little smoke is preferable to using too much. The goal should be to use just enough smoke to keep the bees off balance so the inspection process can proceed smoothly. When too much smoke is used, the bees will tend to run around in circles, seemingly in a state of panic, which makes it much more difficult to locate the queen. By using just enough smoke to keep the bees simply distracted, they will tend to move more slowly and deliberately. By not causing a panic among the *mellifera* masses, spotting the queen will become a lot less

A little smoke can help to clear the area of bees before you go to work prying up the frames so you can lift them out of the hive.

It takes some practice and dexterity to pry up and grasp the end of a frame, while holding the other end of the frame in your other hand. Lift straight up slowly to avoid crushing bees.

like trying to find a needle in a haystack. To make the process easier, some beekeepers have an empty hive body equipped with a queen excluder attached to its bottom into which they shake the bees and force them through the excluder with smoke. If a queen is present, she will be left on top of the excluder and easily located.

Once you know where the queen mother is and have placed her out of harm's way, assembling the appropriate frames needed to make up the nucleus colony goes much faster. The ideal arrangement of frames when creating a nucleus colony will position the frame(s) containing unhatched eggs between combs filled with brood. This will help ensure that any queen cells that are built are located near the center of the cluster of bees warming the brood nest. Although the frames of eggs and brood will often have honey and pollen stored in the upper third of the comb, any frames consisting wholly of honey and pollen that are included in the nuc should sit on the outer edges and sandwich the brood, acting as insulation against chilly nights and positioning nourishment near the brood nest, where the nurse bees can access it easily. After filling the space left in the hive body containing the nuc with additional combs or frames of foundation, several frames crawling with young nurse bees should be shaken into the artificial swarm if the nuc is to be left in the same yard as the mother colony. A puff of smoke blown across the face of both sides of the frame before shaking will help prevent a harsh response from any shaken-up and irritated bees. Once completed, the fledgling colony can be closed up with the entrance reduced and left to its own devices to raise a new queen.

Some beekeepers recommend placing the newly created nuc in the location of the mother hive and moving the original hive still containing the old queen to a new location within the same bee yard.

This is done so that all of the older forag-ing bees in the original hive will return to the queenless nuc that is now sitting where they have always known their hive to be. This will vastly increase the number of bees in the newly made split, greatly increasing the nucleus colony's chances of successfully raising a new queen. The downside of this technique is that it removes the majority of foraging bees from the mother colony and significantly reduces the honey-gathering potential of the hive even more at a time when nectar-gathering opportunities are often peaking. As a result, both the nuc and the mother hive may need feeding to make it through the winter.

This problem can be mitigated by intro-ducing a mated queen into the nuc. However, a mated queen does not need to be purchased and the nuc won't have to be placed in the location of the original hive as long as there is sealed brood placed within the nucleus colony, a fair number of bees are on the frames that are transferred into the split, and several more frames of bees from the brood nest are shaken into the nuc. This will allow for an adequate number of bees within the new hive, allowing the colony to successfully raise a queen and get off to a good start without your having to resort to moving the mother colony and siting the nuc in its place, or moving the colony to a location over two miles away. If any supplemental feed-ing is required, only the newly made nucleus hive will usually need it, rather than both the nuc and the mother colony. (I want to stress again the importance of remembering to reduce the entrance of the nuc so that the hive can be defended by a minimum number of guard bees.)

A typical success rate of at least 80 percent can be expected when making nucs in a single ten-frame hive body during the natural swarming season, using four or five frames from the breeder colony and filling out the remaining space with frames of drawn comb or

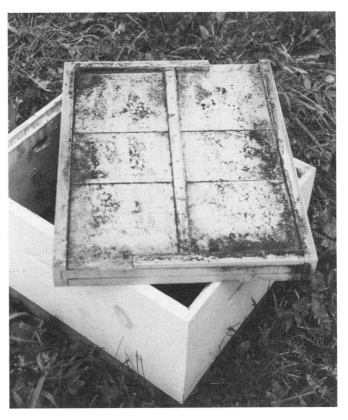

For the double nuc system to work, there must be no way for the bees to be able to move from one side to the other. This mod-ified bottom board helps accomplish this.
Photo by Hilde Whalley.

foundation. The success rate can be boosted well above 90 percent by placing two smaller nucs side by side in a single hive body. The newly created nucleus colonies can be separated by inserting a divider such as a division board feeder with its openings sealed in the center of the hive body. This requires the modification of a bottom board so that it sports a raised section down the middle that is flush with its outer edges and rests against the bottom of the divider, eliminating any openings between the two sides of the chamber. Small openings must be created in the opposite ends of the bottom board, to provide separate entrances for the two hives that will be created. An empty poly grain bag makes an effective inner cover for this double nuc setup. The flexibility of the bag allows for one side of it to be peeled back to expose one nuc within the duplex hive with the residents of the other side being none the wiser. Duct tape can be used to seal the openings in the feeder to prevent commingling of the bees through the shared wall. To give each fledgling nucleus enough room to expand, a frame of foundation can be placed in the outside position farthest from the center of the brood nest.

The benefits of making double nuc colonies are that they make much more efficient use of available equipment, while relying on a minimum of specialized equipment, and maximize the opportunities for success. Between the smaller space that the bees must maintain within the nuc and the fact that the shared wall between the two nascent colonies is kept warm by the fledgling hives on either side during cool weather, stress on the hives is decreased, which allows for greater success in raising a new queen when compared to a single nuc in a ten-frame hive body. This method of nuc making seems especially well adapted for use with Russian strains of bees, which seem to thrive in small spaces. All that is required of the beekeeper is to visit approximately thirty days after making the nucs and transfer each of the mini-hives out of the double nuc box and into its own ten-frame hive body. Positioning yourself with the sun shining over your shoulder should help provide adequate light to see if any eggs have been deposited in the back of the brood cells. On the rare occasion that you cannot find any eggs because one of the nucs has not successfully raised a queen, the frames of bees and honey from the queenless nuc can be combined with the colony that successfully raised a mother. Don't be fooled by the queen that, for some unknown reason, has delayed the start of her egg laying. Every once in a while I have come across a hive that has no eggs present but does have large areas of empty comb with cleaned and polished cells in the center of the frames that the bees are keeping clear of honey and pollen in anticipation of the queen beginning her egg laying. In these cases, leaving the colony alone for another week will provide sufficient time for the queen to start laying.

If made in May or early June here in Vermont, successful nucs will often fill up their hive body and an additional super with enough honey, bees, and brood to get them through the winter in good shape. Nucs made later in the season, however, or those made with foundation rather than drawn comb almost always require additional feeding to prevent starvation from descending upon the young hive before the spring blooms herald the start of the next honey season.

There are a number of different nuc-making boxes that are offered by the bee supply companies. Although they may work perfectly well, I have never tried them, preferring to utilize equipment that I already own rather than buy a new, specialized nuc box. As a result, I have always made nucs in standard ten-frame hive bodies, and occasionally I have used a modified bottom board and divider of some type to separate a hive body into two individual compartments with room enough for the two small nucs.

The big question for those beekeepers who want to increase the numbers of hives in their care through queen rearing or nuc propagation is: Which hives are the best ones to use for breeding purposes? The answer to this question is simplified for those with few hives, because their only option is to breed from the hives that are readily available. However, operations containing large numbers of hives must decide from among their many colonies which of the hives are most worthy of being bred and encouraged to continue their particular genetic lineage. There are many traits that breeding can help bring forth in the bee yard. The typical characteristics that beekeepers have sought to encourage include high honey production, gentle demeanor, low swarming tendencies, fast spring buildup, and conservative honey usage during winter. However, the most important trait during these times of stress on the honey bee population is pest and disease resistance, with *Varroa destructor* by far being the biggest challenge facing the honey industry in the late twentieth and early twenty-first centuries. As a result, the breeding that is currently being

The easiest way to hive a swarm is to put the bees in the hive before *they swarm.*
Photo by Hilde Whalley.

done—by folks like Kirk Webster, Dr. Spivak, and the researchers at the USDA Agricultural Research Service laboratories, along with every backyard beekeeper breeding colonies with the evolution of the bee in mind—is among the most important work going on in apiculture today. The effort to help the bee reach a level of immunity against varroa and other pests and diseases is the long-term solution required to ensure the ultimate survival of the species. All the management techniques and methods that beekeepers use to help the honey bee survive (many of which are described in subsequent chapters) can be seen as stepping-stones—temporary Band-Aids that are applied in an effort to buy the honey bee enough time to do the work required for its evolutionary growth and survival.

The famous German teacher and philosopher Rudolf Steiner certainly touched on this subject over eighty years ago when he noted how the techniques and mind-set of those who care for honey bees affect the evolutionary process and, thus, the ultimate survival of this fascinating insect. As a result, during a series of lectures, Steiner accurately predicted the difficult times the beekeeping industry faces today. Here are Steiner's own words from a lecture he gave on bees on November 26, 1923:

One is able to say—in the whole inter-relationship of the bee-colony—of this organism—nature reveals something very wonderful to us. The bees are subject to forces of Nature which are truly wonderful and of great significance. One cannot but feel shy of fumbling among these forces of Nature. It is becoming increasingly obvious today that wherever man clumsily interferes with these forces he makes matters not better, but worse. He does not make them worse all at once, for it is really so that Nature is everywhere hindered, though notwithstanding these hindrances Nature works as best she may. Certain of these hindrances man can remove, and by doing away with them can make things easier for Nature. For example, he seems actually to be helping Nature when he makes use of beehives which are conveniently arranged, instead of using the old straw skeps.

But here we come to the whole question of artificial bee-keeping. You must not think that I am unable to see—even from a non-anthroposophical point of view—that modern bee-keeping methods seem at first very attractive, for certainly, it makes many things much easier. But the strong holding together—I should like to say—of *one* bee-generation, of one bee-family, will be impaired in the long run.

Speaking generally today, one cannot but praise modern bee-keeping; so long as we see all such precautions observed of which Herr Muller has told us, we must admire them in a certain sense. But we must wait and see how things will be in fifty to eighty years time, for by then certain forces which have hitherto been *organic* in the hive will be mechanized, will become mechanical. It is not possible to bring about that intimate relationship between the colony and a Queen that has been *bought*, which results naturally when a Queen comes into being in the natural way. Only, at first these things are not observed.[2]

Then on December 5, 1923, Herr Muller asked Rudolf Steiner the following question:

I myself, cannot understand that within the next eighty to a hundred years the whole stock of bees will die out. I really cannot understand what Dr. Steiner means by saying that within eighty to a hundred years bee-keeping will be endangered.[3]

To which Steiner answered, in part:

When one makes these experiments one discovers the following:
One finds that calves bred from cows that have been brought to an excessive production of milk, are considerably weaker. You see it the way the remedy affects them. The working or nonworking of the remedy, so to speak, can be tremendously increased in such cases. The calf grows up if it does not die of the disease, but the calf bred from a cow that has been over-stimulated to this over-production of milk, a calf of such breeding is weaker than calves bred from cows that have never been so forced. This change can be observed through the first, second, third, or fourth generations, but is then so slight that observation is not easy. This breeding for milk-production is

89

still of short standing, but I know very well that if it continues, if a cow is forced to yield six gallons of milk a day, if you continue thus maltreating it, all breeding of cows will after a time go absolutely to ruin. There is nothing to be done.

Well, in artificial bee-keeping things are, naturally, not fundamentally so bad, because the bee is a creature that can always help itself again, that is indeed, incredibly able to help itself because the bee lives so much nearer to Nature than the cow that is being bred in this fashion. It is not even quite so bad if cows so maltreated for milk-production are nevertheless at times taken out to pasture. But on the big dairy farms this is no longer done. These farms have nothing but stall-feeding; the cow is completely torn away from natural conditions.

You cannot afford to do this in bee-keeping. Thanks to its nature the bee remains united with external Nature; it helps itself again. And you see, gentlemen, this self-help in the bee-hive is something extremely wonderful.[4]

Our tendency to break things up into component parts rather than working with the entire system leads to the mechanized approach referred to by Steiner. It is important to keep in mind that, while we are dealing with individual honey bees, we must also respond to the hive in its entirety. Each individual honey bee can be likened to an organ or cell within the body of the colony, whereas the entire hive is the individual. Such a holistic vision has long been a primary component of truly organic agriculture.

Notes

1. Jeffery W. Harris and John R. Harbo, "The SMR Trait Explained by Hygienic Behavior of Adult Bees," *American Bee Journal* 145 (May 2005): 430–31.

2. Rudolf Steiner, *Beekeeping: Nine Lectures on Bees,* 2nd printing (Blauvelt, NY: Steinerbooks, 1988), 14–15.

3. Ibid., 40.

4. Ibid., 42–43.

CHAPTER 5

PARASITIC MITES

If skies remain clear, the air warm, and pollen and nectar abound in the flowers, the workers, through a kind of forgetful indulgence, or over-scrupulous prudence perhaps, will for a short time longer endure the importunate, disastrous presence of the males.

—Maurice Maeterlinck

Adequate control of honey bee pests and diseases is recognized as an integral part of good beekeeping practices. Until the introduction of organic standards, the only federal laws affecting honey bees were regulations on importation, developed to reduce the spread of diseases and problem insects, and regulations on pesticide use, designed to reduce unintended bee mortality. State laws governing the interstate movement of bees and the registration of hive locations were also focused primarily on disease prevention. While many regulations were originally designed to control the spread of American foulbrood, the recent introduction of Africanized honey bees (AHB), small hive beetles (SHB), and varroa mites to the United States has increased the importance of such governmental oversight.

As we've discussed, the best approach to addressing disease and pest issues from both the bee's and the beekeeper's perspective is through the propagation of resistant bees. If the bees have the ability to keep themselves healthy and strong, the apiculturist can focus less time and resources on such maintenance issues and instead direct energies toward growth and production concerns. Unfortunately, the development of truly disease- and pest-resistant bees is likely to take many years to accomplish. In the meantime, what is the organically minded beekeeper to do?

Conventional disease and pest control procedures typically have the favorable attribute of significant labor savings over organic approaches. For the cost of a couple dollars or so, a prepared "dose" can be purchased and introduced to a colony quickly, with minimal effort, to achieve the desired result. In general, organic controls are more labor intensive, even when they are comparable in cost. Some organically approved remedies also require additional equipment prior to use, which adds to their expense. Increased labor requirements combined with the additional expenses associated with organic certification result in organic hive products costing more to produce, and therefore commanding a premium price relative to their conventional counterparts.

In general, it is important to incorporate regular colony inspections into one's hive management program. Periodic inspections become even more critical when utilizing organic methods of disease and pest control. Recognizing the signs of disease or pest predation is crucial to timely intervention to prevent colony decline or collapse. The organically minded beekeeper must also learn to differentiate between, and prioritize among, the afflictions affecting a hive in order to choose the appropriate course of action for a given situation.

Efforts to detect diseases and pests begin as soon as one enters the apiary. Simple observations such as the amount and type of activity in front of a hive, or the presence of fecal matter or dead bees lying by the entrance, can reveal much about the condition of a colony before you even remove the outer cover to inspect the hive. Good beekeeping practices demand that the beekeeper remain alert to such clues regarding the hive's condition and learn to recognize their meaning. Working with honey bees teaches us the importance of paying attention to the fine points that are so easily overlooked. This attention to detail plays a crucial role in the survival of honey bee colonies, whether they are managed organically or otherwise. Noticing when the soothing hum that typically emanates from a hive has changed in pitch, when the faintly sweet scent that normally exudes from a hive becomes sour or foul, or when the movements and activities of the bees themselves are not in keeping with a happy and harmonious hive—these are the kinds of small observations that can provide clues to the status of a colony and how to assist the bees as their keeper. Of course, being able to identify problems within the hive requires that we first take the time to observe and experience the way things are when all is right within a colony. Once you become familiar and comfortable around the bees, you may even be able to use your intuition to sense when something is not right with a hive. Anyone who keeps bees long enough is likely to come to the realization that beekeeping is not an exact science. There is as much art as science involved in the practice of this ancient craft. All too often, just when we think we have the bees all figured out, a hive will do the unexpected. Beekeeping is one of those activities in life where the more you know, the more you realize how little you *truly* know.

All too often, just when we think we have the bees all figured out, a hive will do the unexpected. Beekeeping is one of those activities in life where the more you know, the more you realize how little you *truly* know.

When it comes to hive inspection and manipulations, there are two basic philosophies I have heard expounded over the years, and they are diametrically opposed to each other. The first believes that it is best to disturb the bees as little as possible. Honey bees know how to conduct their own affairs, it is argued, and the less disruptive we are of their organization and work, the better. The alternative view calls for regular hive manipulations and inspections, both for early detection of any potential problems and to cause the bees to constantly reorganize and redouble their efforts following each manipulation of the hive, further stimulating honey collection activities.

My preference is to follow the middle road. Striking a happy medium between the two extremes means that one incorporates a single thorough inspection of every hive early in spring, with regular cursory assessments approximately every seven to fourteen days during the rest of the active season. Such quick inspections typically consist of observing the bees at the entrance of the hive and peeking under the inner cover to make sure everything looks normal and that there is plenty of room for additional honey storage. With experience, these observations are all that should be needed to ascertain whether a colony is evolving in a normal and healthy manner. If during a quick check a problem is sensed or signs of trouble are noticed, a more thorough inspection is in order. Typical indications that a hive may be having difficulties and requires a closer look include weak hives that have an insufficient number of worker bees, colonies that sound or act queenless, and hives in which no worker brood is evident as one peeks down in between the frames of the brood chamber. Otherwise, it is generally preferable to stay out of the brood nest and leave the bees to themselves to do what they do best.

Varroa destructor

Observing bees as they go about their business is truly fascinating. For me a large part of this fascination stems from the world honey bees create within the hive and its many similarities to human society. The constant activity and industriousness of the colony, and the clear delineation of individual responsibilities and jobs, with all activity geared toward the singular goal of the common good, leave ample room for awe and inspiration. Unfortunately, observation can also cause worry and concern when disease or parasites such as varroa mites are detected. Spotting varroa within a colony is made easier by the fact that it is much larger than its cousin, the tiny tracheal mite. When you examine a frame of brood, varroa mites may be seen crawling around on the bees or, in cases of heavy infestation, even on the frame itself. However, spotting a mite clinging to a frame or the exterior of a bee that is scampering around on a comb with hundreds of other bees can be a challenge. Good lighting and a sharp eye are necessary for detection. About the size of a pinhead, varroa take on a reddish brown hue as they mature. This deep rustlike color makes the mite relatively easy to see when con-

Often the first chance we have to see the individual varroa mite is on drone brood. Brood exposed when supers are separated will typically be drone brood, which is the mite's favorite breeding ground. Photo by Tad Merrick.

trasted with the white of the developing pupa upon which it feeds prior to emerging from its birthing cell.

Aside from the direct observation of varroa, a telltale sign that mites have infested a hive is the presence of bees with deformed wings. These sickly honey bees can typically be spotted in front of the hive trying in vain to fly, though usually they are weak and seem to be in the last throes of their lives, if they are alive at all. In addition, the hive exhibits a general weakness that expresses itself in deteriorated brood and declining worker populations. These symptoms are collectively known as parasitic mite syndrome (PMS).[1] Viruses that are spread by *V. destructor* are the primary cause of these bee PMS symptoms.

Even if you didn't see the mite on the thorax of her sister nearby, bees with deformed wings are a typical sign that varroa mites are present in the hive.
Photo by Steve Parise.

Although varroa are able to reproduce on worker brood, research has shown that the mites prefer to raise their young on *A. mellifera*'s drone brood. Thus, the regular inspection of drone brood can be a good barometer of the level of varroa within a hive. The easiest way to inspect the drone brood for the presence of varroa mites is to simply observe brood that are exposed in the burr comb that the bees build in the space between the supers and hive bodies. These pieces of comb built outside of the frames typically occur when a honey flow is on and the colony runs out of room to store its bounty. The cells the workers build between the supers are most often the larger drone comb cells. These drone cells are typically ripped apart and the enclosed pupae are exposed when the supers are separated. If mites are present, a quick glance at the exposed brood immediately after lifting the hive body or super will reveal them scampering around on the developing larvae in an effort to get out of the sunlight, like rats abandoning a sinking ship.

If a hive has not been crowded enough to build burr comb between the supers, a frame containing sealed drone brood may be examined to check for the presence of varroa. An ordinary capping scratcher (uncapping fork), inserted at a slight angle through the bullet-shaped caps of a section of sealed drone comb and pulled away from the frame, will carry with it the drone pupae within the cells. This method of detection allows both the brood and the now empty brood cells to be easily inspected for the presence of mites.

Another way of testing for the presence of varroa mites was developed early on and adopted by the beekeeping industry. It is known as the ether roll technique. This process requires a wide mouthed canning jar and lid. Approximately two hundred to three hundred bees are collected in the jar (which roughly translates to about one and a half inches of bees in the bottom of the jar). The bees should be collected from the brood nest, and care should be taken not to accidentally include the queen within the jar. With the lid in place, a light rap of the container with your hand will knock the bees down to the bottom of the jar so a fairly accurate estimate of the number of bees the jar contains can be made. Once an adequate number of bees have been collected, the lid of the jar is cracked open and the honey bees are sprayed

For quick results when checking for varroa infestation, the uncapping fork, or capping scratcher, can be used on areas where sealed drone brood is concentrated.
Photo by Hilde Whalley.

with ether from an aerosol can, such as those typically sold in auto supply stores for use in helping to start engines. A short, two-second blast of ether is enough to cover the bees. The lid is then replaced and the jar is rotated so that the bees roll around on the side of the glass. Most of the mites that are present on the bees will fall off and stick to the side of the jar. Alternatively, ethyl alcohol or methyl alcohol (rubbing alcohol) can be poured into the jar and the bee mixture shaken for a minute or two. When alcohol is used, the bees are separated from the alcohol by pouring them over a wire mesh. The alcohol is then strained and the number of mites examined.

Unfortunately, all the bees within the jar are killed using these methods, although the use of alcohol may preserve the bees for other purposes such as dissecting them to inspect for disease or the presence of tracheal mites. In addition, the use of a toxic engine-starting spray does not really resonate with the natural beekeeping philosophy. An approach that is more in line with the organic state of mind consists of collecting some bees in the manner described above but, instead of adding ether or alcohol, dropping in a tablespoon of confectioners' sugar. The powdered sugar should be fresh and flow freely, with no clumps. Shaking and rolling the sealed jar will coat the bees with sugar and make it difficult for the mites to maintain their grip on their hosts. The lid of the jar can then be replaced with a piece of one-eighth-inch hardware cloth, and the sugar, along with the mites, can be dumped into an empty jar, leaving the bees behind. The lid of the canning jar makes a good template when cutting a piece of hardware cloth to act as a screen for separating the bees from the mites and loose sugar. A fine nylon mesh can then be placed over the mouth of this second container that will allow the sugar to be separated from the mites in order to make the varroa more visible for counting. Alternatively, water can be added to the mite and sugar mixture, which will dissolve the sugar and make the mites visible. The bees, still covered with a dusting of powdered sugar, can be returned to their hive, none the worse for wear.

A less labor-intensive method of determining varroa levels within a hive relies on the fact that a small percentage of mites will naturally fall off their hosts during a given period. This allows for the use of a sticky board trap to catch the varroa that drop to the bottom of the hive. Although I have yet to use a sticky board myself, it has become a common technique for evaluating the level of mite infestation within a hive. Often a sheet of white cardboard or heavy construction paper is used for this purpose. The face of the board is typically coated with oil so that mites coming in contact with its surface will stick to it and be unable to crawl away. (A paint roller soaked in oil works well for applying an even coat.) The sticky board is then placed on the bottom board of the hive. With a one-sixteenth- or one-eighth-inch wire mesh over the sticky board, mites that lose their grip and fall off the bees will fall through the screen and stick to the board, and the bees are prevented from becoming entangled themselves. A single sticky board is best used for a short duration of one to seven days at most. Leaving the board in the hive for longer periods may allow it to collect significant amounts of hive debris, making examination for mites difficult. Lines drawn on the face of the board dividing it into sections about two inches square can greatly assist in counting the mites captured. A magnifying glass and good lighting are also very helpful.

I've provided this information on techniques to evaluate varroa levels so the reader will have some background and knowledge on what is considered to be standard operating procedure within the beekeeping industry. In actual practice, however, many beekeepers, including myself, find that we just don't have the time or inclination to carry out these laborious procedures. I am constantly working to simplify my operation by reducing the amounts and types of equipment I must rely on. Because the regions where bees are isolated enough not to be exposed to mites are few and far between, I simply always assume that my hives always contain mites. By regularly inspecting and being on the lookout for the basic indicators of varroa presence within the hive—mites in burr comb brood, mites on bees, and bees with deformed wings—a good feel for the level of mite infestation within a hive can be obtained without the use of sampling methods. Looking for indications of varroa's presence while performing regular checkups for supering purposes every week or two, between late May and early August, allows me to make sure that the varroa mite levels within my hives are not getting out of hand.

> By keeping the varroa population as low as possible during the time when the hive is raising the majority of bees that will overwinter within the hive—September and October—we can help ensure a healthy honey bee population to carry the hive into spring.

When I see no deformed bees, or I observe few or no mites in burr comb and on bees during inspections, I assume that mites are present but in such numbers that the low-impact passive controls I constantly keep in place are adequate to keep the hive healthy for the time being. When several mites are observed within the brood cells, this is a sign that mite levels are starting to build, and a close eye must be kept on the hive to ensure that the mites do not become too populous too quickly. When bees with deformed wings are observed, then the bees have the double trouble of both viruses and mites weakening their immune systems, and I know emergency high-impact controls are called for in the near future to avoid the collapse of the hive within the next few brood cycles.

Now, the reason none of this sounds very scientific is that it isn't. However, it works for me given the time and resources I have available. Using this basic observational method I have kept my average yearly colony losses to well under 20 percent since 2001, whereas most conventional beekeepers have been reporting regular losses of 40 percent or more. The scientific approach would be to take a count of the number of mites present, either by an ether roll, a sugar roll, or a sticky board test, and to evaluate whether the number of mites present warrants some type of mite control treatment. In actual practice, however, I don't find this to be a very practical approach. Even if we could find the time, most sideline or professional beekeepers will have too many hives and too little time to make consistent monitoring realistic. Regular testing of mite levels would make more sense to me if I were using a mite control method for which the mites are likely to develop genetic resistance. It might also be appropriate if I were extremely concerned about the cost of the additional labor and materials involved with mite treatments and felt that the additional cost of monitoring mite levels outweighed the cost of potentially applying unnecessary treatments. However, because most organic approaches do not use mite treatments that allow for the development of resistant mites, and because I keep the number of hives under my care within a human scale so that the time and money invested in treatments that may be applied prophylactically are not prohibitive, the savings in labor and time from forgoing mite sampling and simply assuming that all hives are infested with varroa and require some form of treatment is worth it for me. By forgoing the purchase of labor-intensive mite-monitoring equipment, my constant goal of simplifying and keeping the amount of beekeeping equipment I rely on to a minimum is more fully realized. This approach could be considered a weakness in my beekeeping philosophy, especially by the large beekeeping operations that must work hard to eliminate all unnecessary labor and input costs to stay profitable. In such cases, large outfits may compromise by monitoring a sampling of say, just twenty-five hives on a regular basis to keep tabs on mite levels among their hives in general.

Up to this point, I have been able to keep my bees quite healthy throughout the seasons here in Vermont without sampling for mite loads by continually using a variety of low-level mechanical controls from early spring through the middle of August. I find that an effective, high-impact active varroa treatment of a nontoxic nature should be applied in late summer, after the honey harvest, to minimize winter losses and help guarantee strong hives in the spring. This means that the work to produce next year's honey harvest actually begins in August, and it requires that I schedule my honey harvest earlier in the season than has traditionally been the case. Early harvesting allows me to treat each hive before the mites reach the level where they will cause the hive to collapse. As a result, I have virtually eliminated the loss of hives to mites during the spring, summer, and fall, limiting any colonies that are killed by varroa to the winter and early spring. In my climate region (zone 4 on the USDA plant hardiness zone map), I find that an early treatment in mid- to late August also helps to ensure that the favorable temperatures necessary for the effective application of many nontoxic controls will be available. If I wait until September in order to maximize the honey crop, there is a greater risk that it will be too cool for some of the temperature-sensitive treatments to be effective, and the hives are likely to collapse from their mite load by midwinter. By keeping the varroa population as low as possible during the time when the hive is raising the majority of bees that will over-winter within the hive—September and October—we can help ensure a healthy honey bee population to carry the hive into spring. An early honey harvest also gives the colony the opportunity to gather additional honey stores and helps guarantee access to plenty of food for the hive over the long winter season. A drawback to this approach is the fact that often the hives become so full of honey in September that late-season swarms may be thrown off by the colony due to the overcrowded conditions. It is disheartening to know that these swarms, if not captured and fed supers of honey that have been harvested from other hives, will not have enough time to gather and store enough honey on their own to make it through the winter.

> From an organic perspective, the effort to eliminate every mite within a colony is misguided. In many instances, the bees can sustain an infestation rate of 10 to 15 percent with little harm. Allowing hives to be exposed to low levels of varroa on a constant basis will speed up the development of the honey bee's natural genetic resistance to the mite.

This early-season approach to mite treatment requires that the beekeeper be willing to sacrifice what can sometimes be a significant portion of the honey harvest. Late August and September is the time when the bees gather rich, dark, and flavorful goldenrod honey in Vermont. Even though the hive may not technically need it, by leaving this additional honey for the colony and treating the hive for varroa, the ability of the hive to survive the winter and find itself with a healthy population of workers come spring is greatly improved. I have also discovered that without the darker, more robust flavor of goldenrod, the honey that is harvested and bottled has the consistently lighter color and more delicate flavor that most of my customers prefer.

Treating for Mites

The best approach for treating varroa will vary depending on the level of infestation within a colony, the time of year, and the resources available. For example, certain high-impact treatments that may be used to prevent the collapse of a hive with high levels of mites may not be possible if honey supers are on the colony. In contrast, relatively few mites within a hive during the start of a major honey flow may be adequately addressed with low-impact mechanical mite control procedures that remove only a small percentage of mites but have no effect on honey that is to be harvested and sold. Conventional wisdom dictates that periodic sampling is required to estimate the population of varroa within a colony and to determine what type of mite control methods to use and when to use them. However, determining which level of infestation requires which type of treatment can be a challenge due to the large number of variables involved. When considering mite levels, it would be much simpler if there were a hard-and-fast number that a beekeeper could use to know when and how to treat for varroa. Some people have attempted to develop just such benchmarks; however, these efforts are complicated by the fact that the same mite populations within different colonies may result in a variety of sampling results, depending on whether a sticky board or one of the roll methods is employed. To complicate matters even more, there is some disagreement over what level of mites is a cause for concern. Are chemotherapy treatments indicated when the level of mites will adversely affect the honey harvest and cause economic damage, or only when the mites reach the point where the hive's survival is at risk? Additionally, the threshold indicating the need for mite treatment will be lower if the hive's immunity is weakened by the presence of other diseases or pests. Ultimately, the beekeeper must make the final decision. The approach that I have evolved over time relies on numerous low-impact, mechanical controls year-round, with a single high-impact, nontoxic control once a year in late summer to reduce the level of mites as much as possible as the hive heads into the winter. This approach has freed me up from having to test my hives for varroa levels at all.

As indicated above, I tend to divide organic varroa treatments into two general categories: active, high-impact, emergency treatments that will remove over 70 percent of mites within a hive during a single treatment and will do so within a short amount of time; and passive, low-impact, mechanical or cultural actions that will affect anywhere from 5 to 30 percent of mites, but that can be used over long periods and are useful in helping to keep mite loads low between high-impact treatments without fear of contaminating honey gathered during major honey flows. Two high-impact treatments that remove the majority of mites in a hive are often required each year, once in spring and once in fall, to ensure the bees do not succumb to mites. However, a single high-impact approach can be utilized with success when low-impact varroa management activities are carried out during the rest of the year, serving to replace one of the high-impact treatments. By using a combination of four or five more low-impact passive mite control activities to replace high-impact treatments, colonies can be kept alive for several years through low-impact mite treatments alone without having to use any major chemotherapy methods.

I have to admit, however, to mixed results relying solely on mechanical and cultural controls, in much the same way that many beekeepers have had limited success utilizing a single high-impact treatment once a year. Until the bees have evolved a higher level of natural resistance, the best method of preventing winterkill from varroa is to use either two high-impact treatments during the year or its equivalent: one high-impact method in late summer, combined with several low-impact controls throughout the rest of the season. Mechanical and cultural controls can be relied upon solely only during the first year that packages or nucleus colonies are managed, as long as they are relatively free from mites to begin with.

Besides having the ability to be utilized while a honey flow is in progress and honey supers are on the hive, low-impact passive varroa controls typically offer the added benefit of being inexpensive. (Some low-impact controls can be time-consuming, however.) Active high-impact varroa controls rely on introducing a foreign substance into the hive, often with the assistance of specialized equipment, and as a result these methods of mite control tend to be more costly.

As mentioned earlier, the number-one threat to honey bees and the beekeeping industry in the late twentieth and early twenty-first centuries has been *Varroa destructor*. The reaction by the U.S. beekeeping industry to the introduction of this mite into North America has been zero tolerance. The chemical controls Miticure, Apistan, and CheckMite were introduced with the expectation of a kill rate of 97 percent or better. From an organic perspective, the effort to eliminate every mite within a colony is misguided. In many instances, the bees can sustain an infestation rate of 10 to 15 percent with little harm. Allowing hives to be exposed to low levels of varroa on a constant basis will speed up the development of the honey bee's natural genetic resistance to the mite. Pesticides, on the other hand, act like a crutch and have the effect of coddling the bees and preventing them from using their own innate defense mechanisms to deal with parasitic invaders. Leaving some mites in the hive on a regular basis is the only way to give the bees a fighting chance of dealing effectively with varroa on their own terms. We should be encouraging the honey bee's natural genetic resistance by working with the bee's evolutionary process, for this is ultimately the best long-term solution to the mite problem.

Organic Varroa Control Measures and Methods

Unfortunately, simply ignoring the mites will, with the rare exception, result in the loss of the hive within one to two years. I therefore recommend that, in addition to starting with naturally mite-resistant stock, organic beekeepers adopt some combination of the following mite control approaches. Although not all of these varroa control techniques have been "scientifically proven" as yet, they have nonetheless been utilized in the field with varying degrees of reported success.

Cultural Management Practices for Varroa

A cultural management practice refers to an activity that is a built-in part of one's hive management system. It can be represented by a piece of equipment that is part of the hive's standard architecture at some point during the season or a hive manipulation that is a regular part of the management cycle carried out by the apiculturist. As such, cultural management practices tend to be physical in nature and act as passive control measures.

Nucleus Colonies

Early on, when varroa first became established in the United States and no treatments had yet been developed, it became obvious that newly made splits from the previous year had a much higher survival rate than the older, stronger hives that had already overwintered a season or two. As a result, one of the low-impact practices that beekeepers should incorporate into their yearly management cycle is the regular production of nucleus colonies as described in chapter 4. This allows the apiculturist to actively expand the number of hives by splitting up those with a proven propensity for survival, increasing the hives' chances of surviving the mite at the same time. Why newly created colonies typically have a better survival rate when faced with varroa infestation than older established hives is a good topic for future research. I believe at least three factors may be at work here. First, by removing bees and brood from a strong hive to make a nuc, we are also removing many of the mites at the same time and interrupting the mite reproductive cycle enough to keep their numbers below the threshold where they will overcome a colony. Second, large families of bees that are represented by strong colonies have larger brood nests to support mite reproduction, and they have many more foragers in the field, which results in increased opportunities for mites to hitch a ride and find their way into the hive. By decreasing the size of the brood nest we reduce the incubation opportunities for the mite. Third, young, vigorous queens in their first year may able to fight off the effects of the mites in a way that their older sisters are unable to match.

The existence of viruses that are carried and spread by the mite is well established. The deformed-wing virus has become a telltale sign of a strong varroa presence. A symptom of parasitic mite syndrome, deformed wings are a typical sign that varroa are lurking somewhere nearby, because the mite has been proven to be the vector that transmits the disease. The presence of this virus is not so much a death sentence as it is a warning that unless something more is done soon, the colony in question is likely to collapse. These days, the health of a hive is strongly tied to the population of varroa found within the colony, and this has led to the establishment of an economic injury level (EIL), which is the level of mites a hive can tolerate without the significant loss of honey production and income for the keeper of the bees. This approach puts the emphasis on the hive as an industrial/economic system whose primary purpose is production, rather than a biological system for which the health and integrity of the bee colony is of primary importance. Clearly the biological perspective is more closely aligned with organic methods and natural ecological agriculture.

Varroa destructor does not have a high reproduction rate and typically lays only between two and six eggs, approximately one every thirty hours during each reproductive cycle.[2] Some 20 percent of mites will enter a second cell and lay again, and, depending on whether the mite has entered a drone brood cell or a worker cell, an average of only about 1.8 adult female mites on worker brood to 2.7 adult female mites on drone brood will manage to reach maturity during the course of each cycle to carry on the process of reproduction.[3] Though few in number, the young female varroa that are able to reach maturity have a high survival rate. This creates an exponential growth pattern in the number of mites found in a hive over the course of a year. Although a hive may start the year off with only a dozen mites in the colony, following the first brood cycle their number will jump to approximately 26 individuals engaged in parasitism. After the next generation of brood hatches, about 57 mites may be found inhabiting the hive. After the emergence of the third brood cycle, well over 125 mites may reside in the hive. The next brood cycle—about three months after the start of the year—may result in over 275 mites present in the hive, and so on. Of course, these numbers all presuppose that foraging workers or drifting drones bring no additional mites into the hive, which, in fact, they typically do. It is this exponential progression, combined with the weakening effect of the young developing mites feeding off the honey bee brood, plus the viruses carried by the mites, that allows the mites in a colony to eventually overwhelm the hive and cause its demise before autumn rolls around, unless the beekeeper, or the honey bees themselves, are able to take steps to keep the varroa population below the threshold that will precipitate the colony's collapse.

Although varroa prefer to raise their young on drone brood, they can be successfully raised on the shorter reproductive cycle of the worker brood. Any varroa egg laid within the standard European honey bee colony within the first twelve days of a worker bee's development, or the first fifteen days of a maturing drone's, will typically have time to develop fully to adulthood. Varroa eggs laid after these times will produce mites that will fail to mature in time and will not survive.[4] Hence, a large population of mites can be raised faster on drone brood than on worker brood, and in general, the more brood in the hive, the more opportunities for the mites to sustain prolonged population growth. Hives governed by a queen with varroa-sensitive hygiene (VSH) genes have the ability to sense which mites are able to successfully reproduce and can interfere with the varroa's reproduction cycle. They accomplish this through hygienic behavior, which motivates the bees to uncap and remove the developing bee larvae that are infested with reproducing mites to a much higher extent than the typical hygienic behavior of other honey bee strains.

> Once colonies tolerant of varroa make up the majority of one's hives, breeding efforts can focus on the development of other traits such as gentleness or high honey production. It doesn't matter how much honey a colony can produce, or how easy it is to work with, if its bees are dead.

Over time, organically minded beekeepers can develop mite-resistant traits in their hives through the regular establishment of nucleus colonies. The simple way to do this is to breed hives that enter the spring with the largest and healthiest bee populations. Hives that can survive and thrive in the presence of varroa should be the primary focus of breeders. Once colonies tolerant of varroa make up the majority of one's hives, breeding efforts can focus on the development of other traits such as gentleness or high honey production. It doesn't matter how much honey a colony can produce, or how easy it is to work with, if its bees are dead from varroa infestation.

Screened Bottom Boards

As an integral part of the hive, the screened bottom board is the one management tool for helping to control varroa mites that requires the least amount of labor. However, as the saying goes, "You get what you pay for," because it provides year-round removal of about only 10 to 20 percent of the mites in a hive. The screened bottom board's effectiveness is based on the observation that a certain percentage of varroa will lose their footing and fall off their honey bee hosts during any given period. When a mite falls off a bee in a hive outfitted with a traditional solid wood bottom, the mite will land on the bottom board and have the opportunity to hitch a ride back up to the brood nest by attaching itself to a passing bee. It is believed that when the mites are over an inch and a half away from honey bee activity, they have a difficult time sensing where the bees are, and thus their ability to find their way back to the brood nest is impaired. When a screened bottom replaces the solid wood bottom, the mites that fall through the screen are in effect permanently removed from the hive.

I have experimented with two types of screened bottom boards. The first design was a screened frame that resembles the basic piece of equipment available through many bee supply companies today. In fact, when the first bottom board screen became available, I purchased one and then used it as a template to manufacture more of my own. This design places the screened bottom on top of the solid wood bottom board so that the surface of the screen is about an inch and a half or more away from the surface of the bottom board below. The screen in effect becomes the bottom of the hive for the bees. The former space that was used by the bees as the entrance to the hive just above the solid bottom board must be blocked off with a piece of wood, which acts as a closure so the bees won't confuse it with their new entrance that now exists between the screen and the first brood box. Often adhesive glue, a thin layer of petroleum jelly, or oil is applied to a catch tray set on the bottom board beneath the screen. This ensures that mites will become stuck after falling through the screen, thus removing any chance that they will somehow be able to find their way back into the hive. A sticky board, similar to those used for detecting and evaluating the presence of varroa within the colony, may also be used for this purpose beneath the screen.

Although the screened bottom above a solid wood bottom board works well in removing fallen mites from the hive, it tends to also accumulate wax and hive debris and requires regular cleaning to maintain the effectiveness of its mite-trapping abilities and to prevent wax moths from laying their eggs in the area. The additional equipment (bottom board screen and closure for the false entrance), the additional labor involved in cleaning out the area between the screen and the bottom board, and changing the sticky board or the sticky material on the catch tray, and the fact that this space below the hive is an attractive home for wax worms and other insects prompted me to remove the solid bottom board. Now all my hives have a screened bottom board open to the ground.

For those who really want to save on equipment handling, cost, and storage, get rid of the bottom altogether. Some beekeepers have reported success by doing just that. However, forgoing the floor of the hive is not practical if you intend to move the hive any significant distance at some point. Because either solid or screened bottom boards can help keep the larger predatory insects and animals out, as well as keeping the bees in, removing the bottom is not recommended if such critters are active in and around your bee yard.

To create a screened bottom board, simply cut out the center of your existing boards and staple a piece of one-eighth-inch hardware cloth, available at your local hardware store, over the opening. Then rejoice in the fact that this solution reduces your equipment requirements for each hive by two items. Now that the bottom of the hive consists of a screen, mites and hive debris that fall to the bottom of the colony will fall through the screen and land on the ground below. Cleaning out the area under the screen is no longer necessary, and even the extremely healthy and active mites that fall through are likely to become a meal for insects that live under the hive, rather than making their way back up into the hive. Replacing the solid wood bottom of the hive with a screen also improves the air circulation within the hive, making it easier for the colony to cool itself during the hot summer months. The increased ventilation also comes in handy when the hive entrances are sealed up, such as when the bees are contained while the colony is being transported or relocated.

Initially I was concerned about how the hives would survive the winter with a screened bottom board. However, I am pleased to report that opening up the bottom of the hive, as opposed to reducing the openings during the winter months, as is often done in northern climates, has not adversely affected my colony's ability to survive during the long, cold Vermont winters. Originally I was worried about the possibility of increased winterkill due to the amplified draft the screen bottom would create. However, my concerns were based on anthropomorphic thinking and have proved to be quite unfounded. Unlike our own winter homes, which must be tightly sealed and heated to keep us comfortable, the bees build their winter accommodations with their bodies by creating a cluster. By snuggling together and using their body heat to warm one another during cold weather, the bees are able to regulate the temperature of their living space without much regard for the size of the opening built into the floor of their abode. Honey bees are similar to humans in one respect, however: if they become wet and are exposed to prolonged cold temperatures, their chances of survival become tenuous. Over the years the bees have proven to me that as long as they have plenty of high-quality honey available and are kept dry, they can typically handle whatever cold temperatures winter wants to throw their way.

One criticism of the screened bottom approach is that, because honey bee brood takes longer to fully mature in cool temperatures, the increased ventilation created by the screened bottom will lower the temperature of the hive and increase the time that the developing varroa mites have to mature within their birthing cells. This in turn, it is argued, helps to actually increase the mite population within the colony. This assumes, however, that the reduced warmth does not similarly slow the speed of varroa development. Although it is possible there may be a slight increase in the number of mites that are able to reach maturity due to the longer length of the honey bee brood cycle, I sense that the screened bottom board removes far more mites than it adds, and it acts as a net gain for the bees at the mites' expense, especially when the benefits of increased ventilation are factored in. However, this would be a good question for researchers to answer definitively through further study.

Another potential drawback of using a screened bottom board is that colonies that are predisposed to using copious amounts of propolis within the hive are likely to try to seal up the opening created by the screen in the bottom of the hive. Over the course of a year, however, I have yet to see a hive propolize more than about a one-quarter-inch border around the outer edge of the screening. However, in warmer climates, where the bees have the opportunity to collect propolis over a longer period, the sealing up of the screen may become more of an issue, and the screen may require cleaning during the course of the year to keep the bottom opening clear.

The major disadvantage of using a screened bottom that is open to the ground is that it can be difficult to use a sticky board beneath the hive should you wish to measure the level of mite drop and evaluate the population of mites within the colony or the effectiveness of an active high-impact treatment. Whether you decide to use a screened bottom over a conventional wooden bottom board or a screened bottom board that is open to the ground, a screen on the bottom should become standard equipment on all hives.

The effectiveness of a screened bottom or screened bottom board in removing varroa mites from the hive can be increased dramatically with the use of various techniques designed to increase natural mite fall. The use of these additional management tools can improve the effectiveness of a bottom board screen, transforming it into an important part of a high-impact emergency control measure rather than just a low-impact tool.

One method that encourages mites to fall and has met with success is the use of smoke. Like all living creatures, mites do not like smoke and drop off the bees in higher numbers when this irritant is applied. Although I have yet to try this approach myself in a systematic and regular manner, smoke from burning the leaves of various plants (tobacco, black walnut, cedar, grapefruit, and creosote bush), as well as mature sumac seedheads, has been reported to be effective in this regard. Depending on where you live, some of these materials may be available to you and some may not; however, pure tobacco that does not have fillers and additives may be obtained nationally through companies such as American Spirit Tobacco Company Research conducted by Dr. Frank A. Eischen of the USDA-ARS Honey Bee Research Laboratory in Weslaco, Texas, has indicated that smoke weakens varroa to the point where they are likely to lose their footing and fall off their bee hosts. This action is brought about through the incomplete combustion that forms the toxic substances and particulates found in smoke. However, it has yet to be established whether specific substances produced by the burning of certain plant materials are toxic to mites, whether smoke itself acts to confuse the mites, or whether the mites fall off simply in an effort to get away from the smoke in much the same way that bees react to smoke. Perhaps all of these actions come into play. It is interesting to note that Dr. Eischen has discovered that some plants, such as the creosote bush, when burned can have a negative effect on the honey bees as well as the mites, whereas others, such as grapefruit leaves, do not seem to harm the bees at all—and yet mites are still observed to drop in high numbers when exposed to the smoke. This observation indicates that it is possible for smoke to act physically to confuse and irritate the mites and at the same time to carry compounds that may potentially act as a parasiticide. It has already been established that smoke from a simple wood fire contains numerous toxic compounds, including dioxin, benzene, and heavy metals.[5] It is not a good idea to treat hives with smoke when honey supers are in place, because heavy smoke can impart an off flavor to the honey stored in the comb, indicating that chemical constituents within the smoke are migrating into the honey.

The application of smoke for mite control consists of filling a standard smoker with burning material and directing copious amounts of smoke into an opening of the hive. Because smoke naturally rises, the most effective approach is to blow the smoke through the bottom entrance of the hive and to close off all other openings that the bees could use for ventilation so that a heavy accumulation of smoke will take place. According to the literature and most testimonial reports, the ideal period for smoke treatments ranges from thirty to sixty seconds. Cool, thick clouds of puffy white smoke is the type required for mite treatments, just as it is when working with bees. When the smoke becomes hot and gray, it is a sign that the smoker needs refueling. Because this control method will not affect mites that are sealed within brood cells or piggybacking on foragers in the field, two or three treatments spaced a week to ten days apart are indicated. In areas where the types of materials listed above (leaves of tobacco, black walnut, and so forth) are common, their regular use as smoker fuel during normal hive inspections and manipulations may offer a degree of varroa control, especially when screened bottom boards are in use.

The biggest drawback to the use of smoke for mite control is the deleterious effect that some of the stronger smokes, like those produced by tobacco and the creosote bush, may have on the bees if they are exposed to too much smoke over too long a period. Issues regarding the effect of smoke applications on the queen and uncapped developing brood, the chance that constituents of the smoke may be absorbed by honey or wax, and the possibility of harm to those who inhale the smoke during its application should also be considered.

Another technique that can be utilized to increase the number of mites that drop to the bottom of the hive and are removed by a bottom board screen is to coat the bees with confectioners' sugar, or some other benign substance such as wheat flour. The use of inert, powdery substances that have a particle size between five and fifteen microns has become known as the Dowda method. This approach is akin to throwing a bunch of marbles underneath each mite's eight little feet and causing them to take a spill. It also increases the bees' grooming behavior. As the honey bees clean the dust off their bodies, mites are more likely to be dislodged in the process.

I first became aware of the benefits of dusting the bees after reading about the experiences of beekeepers who were getting good results controlling varroa using a four-to-one ratio of confectioners' sugar mixed with garlic powder, which was added to help kill bacteria and fungi within the colony. (It is important to note, however, that dusting a hive with a powder will work only when combined with the use of a screened bottom or a screened bottom board.) The typical treatment protocol for an average hive calls for about a quarter pound (four ounces) of confectioners' sugar to be applied evenly to the top bars of the hive. The sugar is then distributed down between the frames with a hive tool or brush, covering the frames and the bees in the process. In actual practice, it is much quicker to just keep shaking sugar between the frames until all the bees seem to be coated. For best results, the hive should be taken apart and each hive body and super treated individually. When treating large numbers of hives, a container such as a plastic five-pound honey jar with holes drilled into the cover makes a great sugar shaker. Also, amounts of sugar needed will vary depending on the bee population, with less sugar required for weak hives and more needed for really strong colonies. Since the goal is to cover every bee in the colony with a light coating of powder, hives with fewer bees require less powdered sugar. As with the use of smoke, mites that are sealed within brood cells or piggybacking on foragers in the field will not be affected. Therefore, hives require regular treatments spaced a week to ten days apart to provide adequate mite control with this method. A number of beekeepers in Florida have reported keeping varroa-infested colonies alive for several years in a row with no other mite treatment than powdered sugar combined with a screened bottom board.[6]

When powdered sugar is used rather than other benign powders like wheat flour or talc, the bees may make use of it as a small source of feed. Because it is unethical and a violation of organic standards to feed sugar to honey bees when it could be stored in honey supers and end up being harvested and sold as honey, the Dowda method should not be utilized at times when honey supers are on the hive. In addition, it is not advisable to dust bees when they could be confined to the hive for extended periods due to cold temperatures, because confectioners' sugar typically contains a small amount of indigestible cornstarch that acts as an anticaking agent and must be eliminated after being ingested by the bees. If powdered sugar treatments must be applied during times when clustering activity is likely, be sure to use pure sugar that does not have an anticaking agent listed as an ingredient so that the risk of dysentery is reduced. You can also make your own powdered sugar simply by processing granulated sugar briefly in a coffee or spice mill. Then you'd be sure there was no indigestible material in the mix.

The use of powdered sugar has great potential as a chemical-free control for varroa, but because small amounts of sugar may be taken in as food by the bees, organic certification committees may be conflicted as to its use. Standards prescribe that all sugar fed to the bees must be from certified organic sources, and at the time of this writing there is no certified organic confectioners' sugar available on the market that I am aware of. Thus, creating your own powdered sugar from organic sugar crystals may be the only way to achieve such control and maintain organic integrity. I suspect that future regulations will be drafted to allow the use of nonorganic powdered sugar for mite control as long as it is not used immediately prior to the application of, or during the presence of, honey supers on the hive. This could also be a great opportunity for some enterprising company to corner the market and start producing and selling organically certified confectioners' sugar.

Traps

The use of traps to capture creatures has a long tradition in human history. Mechanical in nature, trapping is unlikely to allow for the development of resistance in the traditional sense. I suppose that some resistance may develop when trapping higher-thinking organisms, in that the population targeted for trapping may be reduced to those that are smart enough to learn how to avoid a trap. However, the lack of pressure for the development of genetic resistance in *V. destructor* makes trapping a very attractive option to the beekeeper concerned with sustainability in the battle against mites.

In an article in *Science News* from 1989, French researchers reported the results of a study indicating that when mating, the female varroa mite found the European honey bee male drone brood much more attractive than female worker brood. According to the authors of the research, the female mite relies on the pheromones given off by the developing larvae to guide her to the brood cell, and drone larvae produce more of this attractant than the worker larvae do.[7] Once in the cell, the mite feeds on the developing bee and then lays her eggs. The knowledge of this preference shown by the female varroa mite has been instrumental in the development of trapping devices.

The earliest trapping mechanism for mites consisted of a brood frame fitted with a sheet of large-cell foundation that the bees drew out into drone comb. Inserting a frame of drone comb into the hive's brood nest entices the queen to fill the comb with unfertilized eggs. As the eggs hatch and grow into drone larvae, varroa within the hive are attracted to the developing brood. Approximately twenty-six to thirty days after the comb is first introduced to the hive, the frame of sealed brood can be removed along with the reproducing varroa contained within the drone cells. Placing this frame in a freezer for twenty-four hours will kill the mites and brood without damaging the comb. Reinserting the frame back into the hive after it has thawed will then give the bees the opportunity to clean out the cells and allow the trap to be reused. (Some beekeepers will have a spare frame of drone comb that they can immediately exchange with the frame full of brood.) Even though the development period for drones is only twenty-four days, a maximum thirty-day interval for replacing combs is recommended to allow the workers time to clean out the drone cells and provide the queen with time to lay her eggs. An alternative approach to freezing that has been reported to be successful is to direct a forceful spray of water from a garden hose at the sealed comb, making sure that the water strikes the comb at a slight angle. The water will shear off the cappings and then flush away the pupa along with the developing mites.

Although I have not read any scientific research in this regard, I believe that varroa traps are likely to attract only the mites that are found within the general vicinity of the trap's location. As such, it is advisable to periodically vary the location of the trap in order to cover as much of the hive in as equal a manner as possible. For example, if a drone brood trap is placed on the left side of the hive body in position 3 (between frames 2 and 4), a refreshed trap, if used, should be placed in position 7 (between frames 6 and 8). If the brood nest consists of two hive bodies, then one trap should go in the lower hive body while the other trap is positioned in the upper chamber, each on opposite sides of the hive, to try to draw in mites from as wide an area within the colony as possible. Placing traps next to frames of uncapped worker cells that contain eggs will help keep the mites from infesting the worker brood.

Regular rotation of the position of the traps within your hives will increase their effectiveness.

A primary difficulty with using frames of drone comb as a trapping device and then freezing them is that the beekeeper must have access to freezer space large enough to handle the frames. For those with numerous hives, this can prove to be a real challenge, and it is complicated by the fact that storing the frames outside the freezer for any length of time in a warm environment can encourage wax worms or small hive beetles.

Organic beekeeper Joshua White of Northwoods Apiaries in Westfield, Vermont, uses a modification of this technique to trap varroa. Joshua places a shorter frame, such as those found in medium or shallow supers, into a hive body containing the brood nest made up of deep frames. Faced with an abundance of space between the bottom of this short frame and the bottom of the hive body, the bees will naturally fill it with a large section of comb. More often than not, this comb tends to be composed primarily of drone-sized cells into which the queen will lay eggs. When this drone brood is sealed, the excess comb is cut from the bottom of the short frame and the wax rendered. Not only does this remove the mites that are developing within the drone brood and provide the beekeeper with a source of wax, but also the simplicity of a technique that does not require specialized or additional equipment appeals to my sense of Yankee frugality.

One drawback, however, is that the beekeeper is unable to evaluate the immediate effectiveness of this approach. Other than picking the dead brood out of each cell in the frame of drone comb and counting the mites present, there is no way to tell how many varroa have been caught within the drone brood and how effectively the trap is performing. In addition, this method results in the destruction of significant amounts of drone brood. This can be of particular concern for those involved with queen rearing, who are interested in flooding an area with drones available for mating rather than reducing the number of males. As a steward of my hives, I always try to base my management decisions on what is in the best interest of the bees and how I can boost their health and overall immunity, thereby making them stronger. I find that I am personally uncomfortable deliberately killing bees, be they drones or otherwise. Even though they do not directly produce honey or seem to contribute in any active way to the day-to-day activity of the colony, drones are an essential part of a healthy hive.

I don't know how it is for hives down South, where winter barely makes its presence felt, but here in the Northeast winter is not a good season for the drone. For most of the year the drone leads a happy-go-lucky lifestyle when viewed from the human perspective. He does no work whatsoever within the hive, acting as if he were too sick to work and paying no attention to the abundance of chores that require the attention of the colony on a daily basis. With no stinger, the drone does not concern himself with the defense of the nest, preferring to spend his summer days drifting from hive to hive, begging the ever-active females of his species to feed him. Occasionally he'll get up off the couch and fetch his meals himself. The one task the colony calls upon the drone to perform is the act of mating with the queen. This is the one occasion when the drone really shines and rises to the call of duty. This may sound like the males of the bee world have an idyllic, carefree life. However, their fortunes turn with the seasons.

As the cooler days of autumn herald the coming of winter, the workers' instinctive frugality and lack of tolerance for the drones' malingering become evident. They seem to sense that the abundance of nectar-bearing blossoms will not return anytime soon, and that a long dearth of honey will befall the colony during the frozen months ahead. Practical gals as they are, they realize they have to make the most of the honey they have gathered during the spring, summer, and fall, stretching it out until the following spring. At the same time, they recognize the drain on their collective efforts that the drones represent—and thus the worker bees can be seen every October preventing the males from gaining access to their honey stores, driving the drones from the hive, and blocking their reentry. Between the lack of food and the energy expended in vain trying to resist the female workers from throwing them out, the poor drone in its weakened state gets tossed out into the first snow of the season, only to land on its back and buzz dejectedly as if to say, "But babe, what about all the good times we had?" Unfortunately, the drones, like all honey bees, cannot survive for long on their own outside the hive, especially in cool weather. Thus, the lesson we can learn from the drones is to consistently carry our own weight and always contribute in whatever way we can to our family and community, so that we are not looked upon as expendable dead weight during difficult times.

> By snuggling together and using their body heat to warm one another during cold weather, the bees are able to regulate the temperature of their living space without much regard for the size of the opening built into the floor of their abode.

Okay, so perhaps as a man, I'm a little overly sensitive to the destiny of the males within this female-dominated society. Despite the ultimate fate of these bees that have no obvious function other than to mate with the queen, I found that I had an aversion to intentionally killing drone brood. This got me to thinking of alternatives to the drone brood trap. What was needed was a trap design that would capture mites without harming the male bees. To accomplish this most effectively, the trap would need to be situated within the brood nest where the majority of mites in the hive are located. With these basic criteria in mind, and some insightful suggestions from Bill Mraz, I came up with a frame-shaped design with slotted sides that replaces one of the frames in the brood area. The slots in the sides of the frame are large enough for a mite to fit through but too small to allow the honey bee to pass. Positioning the entrance slits vertically allows the mites that enter to make their way down to the bottom of the apparatus unhindered. A strip of paper strategically placed on the floor of the trap and coated with a sticky substance allows the mites to "check in" but not to leave. This form of mite trap has the same localized efficacy as the drone brood trap. It should be staggered throughout the hive at various intervals as described above for maximum effectiveness.

A trap that utilizes sticky paper is the alternative to sacrificing drones in the quest to trap varroa.

The sticky paper within the trap can be made attractive to mites by adding methyl palmitate to its surface at the bottom of the trap. Methyl palmitate is the pheromone given off by the developing drone larvae and was the substance found to be most appealing to female mites in the French studies mentioned above.[8] It can be purchased from chemical supply companies. At temperatures above 85°F (29.4°C), however, methyl palmitate crystals will melt. By folding the sticky paper so that it will hold the pheromone in its liquid form, the desired results may be obtained. The paper that sits in the bottom of the trap should also be folded in such a way as to ensure that the edges are flush with the sides of the trap, so that there is no room for mites to crawl underneath the paper rather than onto its sticky surface.

A significant benefit that a sticky trap offers over a drone brood trap is that the number of mites caught within the trap is readily visible. This makes the sticky paper mite trap very effective as a monitoring device. Because methyl palmitate is naturally present within a colony, is approved by the USDA as an animal feed additive, and is used by industry to reduce foaming during cardboard manufacturing, trapping with pheromone bait should be able to be implemented while honey supers are on the hive without the risk that toxic compounds will contaminate the crop.

Unfortunately, the bees have a tendency to fill in the slits in the trap with propolis when the trap is left in a hive for long periods. However, cleaning the slots is simply a matter of sliding a hive tool along the propolis-filled openings to dislodge any buildup that may have accumulated. This job is made easier when the trap is exposed to cold temperatures first, causing the propolis to become hard and brittle.

When I first began using this method, I removed the traps, cleaning and replacing them with a fresh trap, once every thirty days or so, similar to the schedule for removing drone comb. However, I found this to be unnecessary and extremely time-consuming, especially as the number of hives in my bee yards grew, and more and more of them survived through the winter. I eventually began changing the trap only once, about halfway through the active season, installing the first trap in early May, replacing it with a fresh trap and bait strip around the end of June, and then removing it in August. This

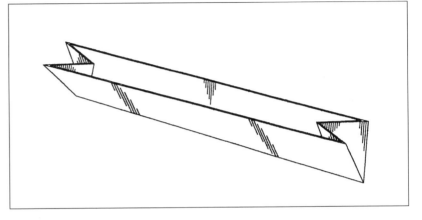

The sticky paper trap allows you to monitor the mite load while at the same time reducing mite populations within your hives. Be sure to fold up the ends if you plan on using methyl palmitate as bait.

has worked well for me, and I believe it is the ideal schedule. Of course there have been years when the traps were never removed, changed, or cleaned until August, during or just after the honey harvest, and just prior to applying mite treatments and tucking the bees in for the winter. When I don't have time to clean out the traps during the season, 15 to 20 percent of my hives will propolize well over half of the openings in the trap. This decreases the trap's effectiveness, but I rationalize that it is the trade-off for eliminating the additional labor of cleaning and replacing the traps during the course of the year. Because a number of other low-impact mite control techniques are utilized throughout the season in my hives, I have been able to get away with using the traps in this less efficient manner. If I did not have other effective mite control measures in use simultaneously, I would be much more diligent in refreshing the traps within the hives on a regular basis.

As you can see from the diagram on page 120, these traps are constructed utilizing a standard top bar combined with a fabricated

19¹/₁₆ inches

17¾ inches

9¹/₈ inches

This trap is easy to make with basic carpentry tools and skills.

bottom bar and end bars, so that when the slotted sides are attached the total width of the trap is one-quarter inch thinner than a standard frame. Old-fashioned metal spacers on the end of the trap provide structural support and ensure that the final dimensions closely resemble those of a standard deep frame, thereby preserving the integrity of the bee space within the colony (see the photo on page 118). Unfortunately, these metal spacers are becoming difficult to find as more and more bee supply companies discontinue their manufacture and availability. However, I'm sure someone with greater carpentry skills than I can come up with a design that precludes the need for the metal spacer and yet maintains structural integrity and proper spacing.

Another drawback of this trap design that I have not had time to improve upon is the use of Masonite board in constructing the sides of the trap. The small particles that make up the Masonite become exposed upon cutting the slots, and as a result, the small pieces that are exposed along the ragged edges of the slits are chewed off by the bees and end up stuck to the paper in the bottom of the trap. Although initially this reduces the effectiveness of the trap, I have found that the accumulation of debris lessens dramatically after the first season, to the point where it eventually ceases to be an issue. Sanding these areas down after they have been cut may help reduce this tendency, but I have not yet had the time or patience to try it.

In 1999 I had the idea of applying for a patent for this mite trap and starting a business manufacturing and distributing it. However, after noticing how much more effective the trap was in catching *V. destructor* when methyl palmitate was used as bait, I sought to get a licensing agreement to market the trap, with the pheromone lure as an integral part of the trapping device. Shortly after starting the process to get methyl palmitate approved by the FDA for use as a honey bee miticide, I sent a letter to the French institute where the researchers had discovered methyl palmitate's properties and requested permission to use their intellectual property, for a fee of course. A couple months passed, and after not hearing back from them I wrote again, this time in English. Again I received no reply. Since the effectiveness of the trap is increased dramatically with the use of the pheromone, it did not seem wise to try to build a business promoting a trapping device without it, so I dropped the patenting and marketing idea and instead am satisfying myself with the opportunity to pass along the design of this device so that do-it-yourselfers have the chance to make use of it. Methyl palmitate may be used as bait within such a trap without infringing upon intellectual property rights, or the FDA's pesticide regulations, as long as the honey harvested is not sold but kept for personal use and you are not selling the pheromone to others as a mite control agent.

One argument that has been made against the idea of using drone brood traps, or traps that utilize the pheromones given off by drone brood as an attractant, is that by consistently removing mites that have a disposition for reproducing on developing drones, are we not inadvertently selecting for mites that will reproduce on worker brood? At first glance this sounds like a valid concern. However, due to the shorter brood cycle of the worker bee as compared to the drone, the reduction in mites successfully able to reproduce in worker cells could have a significant effect on their overall population buildup within the hive. So far, all efforts to remedy the varroa mite problem through a genetic solution has been focused on the honey bee. If drone brood trapping places selective pressure on the mite by increasing the number of mites that prefer to reproduce on faster-developing workers, such an event could serve to improve upon the ability of the mite and the European honey bee to coexist without the population of mites, and their attending diseases, overwhelming the host colony.

Whether or not it selects for mites that prefer to reproduce on worker brood, trapping can be an effective way to remove significant numbers of mites from a hive, and because they work physically, traps can be utilized throughout the active season without fear of contaminating the harvest. Although traps alone have not proven to be enough to keep a colony from crashing due to an overabundance of varroa, when used effectively and in conjunction with other low-level control devices such as the bees natural mite resistance and a screened bottom, they can keep the level of mites low enough throughout the spring and summer that a single high-impact treatment in the fall is all that is needed to ensure that the bees will survive the winter.

Heat

There's nothing like a steamy, sweltering day to help shift the honey flow into high gear. In fact, there are certain plants whose blossoms seem to produce a significant amount of nectar only when the hazy, hot, and humid days of summer kick in. Yet while bees may benefit from increased nectar flows during hot weather, there are drawbacks to such warmth. For one, temperatures above approximately 97°F (36°C) may cause bee brood to overheat and die. And although exposure to temperatures around 119°F (48°C) for short periods will kill mites without harming adult bees, prolonged exposure to such heat will kill bees as well.

On hot days, many bees become occupied with maintaining the temperature of the brood nest within the hive at approximately 95°F (35°C) and controlling the humidity level. Watching the bees line up in the same direction and start fanning their wings to induce a draft to regulate the temperature and humidity of the brood nest is fascinating. After many years as a beekeeper, I still enjoy feeling the warm breeze on my skin as I marvel at the hive's cooperative and efficient air-conditioning procedures. For those who are comfortable around bees, it is a joy to put your nose down by the hive entrance and breathe in the sweet smell that rides the air currents generated by the fanning bees. A deep feeling of contentment runs through me whenever I inhale that fragrant scent that indicates all is well within a colony.

The more individual bees are involved with providing air circulation to the colony, the fewer bees are available to forage in the field. Because of this, numerous schemes have been hatched over the years to provide additional ventilation during hot weather, from vented hive covers to solar-powered fans that provide forced-air ventilation. Always one to prefer the simplest route whenever possible, I find that creating a three-quarter-inch-diameter upper entrance hole in the hive body will do much to improve the movement of air through the hive. Just be sure that any additional entrances that are drilled into the equipment are located below or to the side of the handhold so that the entrance is not inadvertently blocked when moving the equipment, thus increasing the risk of raising the ire of the guard bees.

Additionally, moving the outer cover back a half inch so the forward lip of the telescoping cover sits on top of the front of the inner cover seems to provide enough ventilation up through the inner cover for all but the most extreme temperatures.

Some beekeepers like to shift an entire super or hive body back about an inch to improve airflow and help a colony deal with hot, humid temperatures. I don't like to use this approach because it allows rain to get inside the hives and will invite robbing in all but the strongest colonies. If additional cooling is required, the apiculturist can simply provide the bees with the resources they need to handle the situation themselves, by making sure they have access to plenty of clean water. The workers will forage for water and bring it into the hive, where it reduces the temperature within the colony as it evaporates.

One of the most basic ways beekeepers are able to influence the temperatures inside the hive is through the color they paint the outside of their equipment. White and other light colors help reflect sunlight and keep the interior temperature of the hive cooler. Thus, an extremely low-tech

Cocking the outer cover is a quick, easy, and low-cost method of providing additional hive ventilation.

way to take advantage of the lethal effect of heat on mites is to simply paint the hives a dark color and place them where they will receive lots of direct sunlight. The preoccupation of numerous honey bees with ventilating such a hive, and the subsequent loss of honey production, may well be worth the effect of inhibiting varroa through the increased temperature. Although it is unlikely that a hive painted black and sitting in the sun on a hot day will approach the lethal temperatures needed to directly kill mites, such a hive when used in combination with a screened bottom board may well remove a significant number of mites that lose their footing due to their weakened state of heat exhaustion.

When the hive bodies that make up the hives in an apiary are painted different colors, the ability of the individual bees to identify their colony among others is greatly improved. Paint should be used only on the parts of a hive that are exposed to the elements, as the honey bees' interior decorating preference is bare wood and beeswax, both coated with a very thin layer of propolis. Because bees give off heat and moisture, the exterior of hives should not be painted with oil-based paint, as this will tend to crack and peel over time. Latex paints not only are less toxic and better for the environment than oil-based paints but will breathe and allow for the passage of moisture while at the same time repelling water and its damaging effects. As a result, latex paint holds up much better than any other wood protection I have tried. It is interesting that the bee equipment I painted over ten years ago still looks great—aside from some dirt and scuff marks—even though I used no primer and only a single heavy coat of exterior-grade latex. This is in contrast to buildings, which are typically primed before being painted and end up having to be repainted every four to five years or so due to peeling, cracking paint. It makes me wonder: If bucking conventional wisdom and forgoing the use of a primer coat on hives works so well, how well would it work on my home? One of these days I hope to get adventurous enough to try it and find out.

The observation that mature bees can tolerate high temperatures better than varroa mites has led to the creation of techniques for artificially heating up the honey bees' environment in order to adversely affect the mites. One such technique has been developed and studied by Vermont beekeeper Jeff Cunningham as part of a Sustainable Agriculture Research and Education (SARE) grant project titled "Evaluating a Heat-therapeutic Control of the Honey Bee Mite *Varroa destructor.*" Jeff built a heating apparatus based on one he observed while visiting beekeepers in the former Soviet Republic of Uzbekistan, and he presented the results of his testing at a meeting of the Vermont Beekeepers Association. The unit consists of a portable wooden cabinet, a heating unit, and a wire mesh cage that is filled with bees and placed within the upper portion of the cabinet. Plexiglas windows on the top and sides of the cabinet provide means of observation and ventilation. Honey bees are transferred from their hive by shaking, brushing, or blowing them into a large funnel that facilitates the transfer of bees into the cage of the heating unit, much in the same way that packages of bees are prepared. The process is made easier by performing this task at night, when the bees are not apt to fly as much. Jeff also found that cool weather made the job of transferring supers of bees into the heating unit easier. Also, spring is an easier time to treat with heat, because healthy hives tend to be at their lowest population right after winter. By observing an interior-mounted thermometer and the number of mites falling to the bottom of the apparatus while the interior of the heating unit is warmed up, an accurate and practical indication of the effective temperature range can be established.

Jeff's study indicated that a temperature of 116.6°F (47°C) for up to fifteen minutes was ideal in causing the mites to drop off bees without causing any noticeable harm to the bees themselves. A key factor that affected the unit's effectiveness was the number of bees within the cage during treatment. Jeff found that the bees need to be able to space themselves out to allow ample room for the circulation of heated air. If the cage was overfilled with bees, the bees would form clumps within the heating chamber, and far fewer mites would fall than when treating loosely distributed honey bees

The heat source that Jeff utilizes is a 1500-watt "utility" or "milk house" electric heater with a built-in fan such as those found in most hardware and home supply stores. A generator of some type is necessary for applying heat treatments to apiaries that are not located close to homes or other electrical power sources.

Due to all the preparations involved, which include caging the queen temporarily for safekeeping, and the relatively long treatment time needed for each hive, a significant investment of labor is required (approximately thirty minutes per hive) when using this heating method. The upside is that the financial investment is very low over the long run because the primary cost is in the initial purchase of the materials to build and outfit the heating cabinet and wire cage. Ultimately, this approach is unlikely to be economical to large commercial operations, but it is an attractive option for hobbyists and sideline operations running up to a few dozen hives. The biggest benefit of using heat treatments for varroa control is that the chances that mites will build up a tolerance to the treatment

A small trailer is handy for transporting the heating chamber and associated equipment to and from the apiary. Photo by Charles A. Parent III.

After the hive's population of workers and drones is placed inside, the cylindrical basket is inserted into the heating apparatus to apply the precise mite-killing temperature, for a specific period of time. Photo by Charles A. Parent III.

are extremely low, and there is absolutely no opportunity for contamination of honey or wax, making this option available for use at any time during the season.

Grease Patties

When *Varroa destructor* first appeared in Vermont in the early 1990s, there were no established options for dealing with this new threat. As a result, I started looking for clues that might lead to successful treatment protocols that would help keep the bees alive. I found myself scanning articles in bee journals and eavesdropping on conversations at bee meetings, always on the lookout for any mention of what the beekeepers who had already lived with varroa for several years were doing to counter the mite's damage.

One of the first references to a potential treatment I came across in my search was the observation by some beekeepers that grease patties seemed to be helping hives weather the onslaught of varroa. Traditionally, grease patties have been used to introduce antibiotics into the hive for the treatment and prevention of American foulbrood (AFB). Unfortunately, introducing antibiotics into the hive in this manner has only accelerated the emergence of antibiotic-resistant strains of AFB.

What, then, might be causing grease patties to have a positive effect on bees living with varroa infestations? It has been theorized that the grease from the patties, when it is distributed throughout the hive and covers the bees with a light coating, adversely affects the mobility of the varroa in much the same way it affects the tracheal mite. Unfortunately, because varroa mites are so much larger than tracheal mites, the ability of grease patties to reduce the mobility of varroa is less severe and therefore not as effective. However, when it is combined with essential oils rather than antibiotics, it is logical to assume that the grease patty could prove to be effective for varroa control, while at the same time helping to hinder the mobility of tracheal mites. Essential oils that are toxic to mites will increase the

13½ inches

9½ inches

14½ inches

17 inches

24 inches

36 inches

4x4 or bigger
mesh shelf

¼ Plywood
Sheathing

12 inches

24 inches

16½ inches

Sliding Plexiglass Window

Heat treatment apparatus: dimensions of cabinet with cage in place.

> **A Grease Patty Recipe**
>
> Combine 2 parts saturated fat (Crisco vegetable oil) with 1 part white sugar and thoroughly mix. (Small amounts of honey or essential oil may also be added.) Form this mixture into uniform patties of about a quarter pound (four ounces) apiece; store between sheets of wax paper so that they do not stick together. To use, place a single patty on the top bars of the brood chamber.
>
> An easier application method when treating large numbers of hives is to simply take a spoon and place a dollop of sweet grease equaling approximately a quarter pound (four ounces) on the top bars of each brood nest. Although this method is not as precise as using preformed and preweighed patties, it is less work and much faster.

percentage of mites that lose their grip and fall to the bottom of the hive; therefore, the use of a screened bottom or a screened bottom board is indicated in conjunction with oil-laced grease patties.

Soft Chemical Treatments for Varroa

The use of soft chemicals is the kinder and gentler offshoot of the industrial business model that looks for the "magic bullet" that, when added to the hive management protocol, solves the problem of the mite. Soft chemicals are much less toxic than hard chemicals like coumaphos, yet they are still able to get the job done. The use of chemicals within the hive, no matter how safe they are for humans or other life-forms, is really just a temporary solution, a stepping-stone to help us make the difficult transition from bees that are completely vulnerable to those that can hold their own and even thrive in the face of varroa. Like their hard-chemical cousins, soft chemicals are inputs that come from outside the apiary, and thus we require the efforts of others to make them available. They also require additional labor and resources on the part of the apiculturist. This goes against the basic philosophy of self-reliance and the closed-loop system of farming that is at the heart of the organic approach. Such reliance on chemicals can generally be considered unsustainable in the long run, unless the chemicals can be produced on the farm, or unless their use has the potential to eventually bring such health, balance, and vitality to the hive that eventually chemicals are no longer needed.

Soft chemicals are often used in organic agriculture because they fit with the mind-set of the conventional farmer turned organic—the idea that applying a liquid or powder will solve the problem through the miracle of modern chemistry. Many organic farmers do have a conventional farming background, so this mind-set is common and understandable. Soft chemicals also allow those producers with no other alternative to move in the direction of a more sustainable approach and lessen their reliance on toxic chemotherapy management in the long run.

Essential Oils

An essential oil is the oil of a plant or flower that contains the "essence" of the plant. It will typically smell like the plant or flower from which it is derived. Because it takes large amounts of plant material to extract small quantities of oil, essential oils are very potent. Herbalists have traditionally made use of these oils in their treatment of disease conditions. A number of essential oils kill bacteria, fungi, microbes, and other pathogens. As such, it is not too big a jump to think that oils derived from flowers and herbs that are used to treat human ailments could be used to treat diseases of the beehive as well.

From my experience as an organic grower, I am familiar with the use of dormant oil (oil with no noticeable scent or therapeutic qualities) as an effective treatment for apple tree mites. The oil kills the mites by clogging their breathing tubes and causing suffocation. In the mid- to late 1990s, I began to come across the occasional bee journal article or letter to the editor reporting people's experiments for controlling varroa with essential oils derived from eucalyptus, thyme, and members of the mint family (Lamiaceae). In fact, some beekeepers were even claiming that bees that foraged on peppermint blossoms were benefiting from the minute amounts of peppermint oil they brought back into the hive in the nectar.

Initially I soaked a paper towel in a mixture of eucalyptus and thyme oil, wadded it up, and tossed it into my hives just above the brood chamber. As the essential oils evaporated and affected the mites, the bees would proceed to chew up the towel and lug the pieces out of the hive, helping to further spread the essential oils throughout the colony in the process. Thanks to the bees' efforts, this method of introducing essential oils to the hive had the added benefit of eliminating a second trip to the bee yard to remove the paper towel. The drawback to using paper towels was that the length of time the oils were present in the hive was inconsistent, with strong hives removing the oily towels much more rapidly than weaker colonies. In addition, the mixture of oils that I used varied in effectiveness, as I had no scientific research and testing available on which to base the oil mixture to obtain optimum results.

After reading an article in a bee journal about research being conducted with essential oils absorbed in a spongelike material used in floral arrangements, I obtained a block of Oasis Floral Foam from my local florist. This Oasis material has the ability to absorb a large amount of liquid and is similar to the material described in the research article. The florist sticks the flower stems into the sponge and adds water, so that the sponge acts as both an anchor for the bouquet and a reservoir for water to keep the cut flowers fresh. As was discussed in the article, I cut wafers from the sponge block, soaked them in my essential oil mixture, and laid them on the frames in my hives just above the brood nest. The bees chewed these up as well, but they lasted significantly longer than the paper towels.

For several years I continued to play around with using paper towels, Oasis wafers, and grease patties, all laced with thyme and eucalyptus oil, in my yearly management cycle. Not only did this have some effect on both the varroa and the tracheal mites, but the patties made from grease and sugar would sometimes mean the difference between a hive starving over the winter or surviving into the spring. On a number of occasions, after popping open an inner cover of a hive while unpacking it in early spring, I would find a small cluster of greasy bees living off the patty as they anxiously awaited the return of nectar-bearing blossoms.

Colony winter survival rates improved with the use of essential oils. Unfortunately, although my efforts with essential oils seemed to help some, none of my efforts were successful enough in reducing mite numbers to consistently ensure the survival of the majority of colonies.

Then came published research indicating that thymol, a component of thyme oil found in some foods and considered safe for humans, is extremely deadly to varroa but not to bees or people. I purchased some thymol and started to add it to my essential oil mixtures, which up until that point had been based on intuition and guesswork, rather than anything scientific. Luckily, I didn't have to continue floundering along this way much longer.

Api Life VAR is a thymol-based acaricide (a pesticide that kills mites) that is manufactured in Italy and is likely the final product of that early research on essential oils that I read about. Now available in the United States, Api Life utilizes a mixture of approximately 74 percent thymol, 16 percent eucalyptus oil, 3.7 percent L-menthol, and 6.3 percent camphor, all soaked into a foam wafer that looks suspiciously similar to the Oasis material used by florists. Distributed in the United States by Brushy Mountain Bee Farm, Api Life has been shown in controlled studies to reduce varroa levels by as much as 95 percent within the hive for a high-impact effect.

Each of the ingredients that make up Api Life is approved for use individually in organic agriculture. The primary active ingredient, thymol, is used in food products and generally is considered safe. As the essential oil mixture evaporates from the wafer, the vapors travel throughout the hive. Because the mite-killing compounds are delivered primarily in gaseous form, the opportunity for the comb to absorb residues is reduced, when compared to treatments that spread the active ingredient throughout the colony in solid or liquid form.

The rate at which essential oils evaporate into the air varies with temperature, and this has a great impact on an oil's effectiveness as a mite control mechanism. According to the package directions, Api Life is most effective when applied when average daily temperatures are in the range of about 60° to 70°F (16° to 21°C). The use of Api Life in temperatures around 90°F (32°C) or higher causes the queen to stop laying eggs temporarily, and it may increase the potential for bee or brood mortality. A single-treatment protocol calls for the breaking of a single wafer into several equal-sized pieces and placing them on the top bars in the corners of the brood nest near the outer edge of the brood area. After seven to ten days, the wafer pieces are replaced with fresh ones, and a third dose is applied after another seven to ten days and left in for up to twelve days.

As with all foreign objects placed within the hive, the bees will naturally want to remove these strange essential-oil-soaked foam tablets. To prevent this, Api Life VAR may be enclosed in an envelope of wire screen or mesh. I find it helpful to staple down the corners of the screening to the top bars of the frames to keep the bees from gaining access to the acaricide within the wire envelope. This pesticide treatment should not be used when the honey flow is on and honey supers are on the hive.

Although the instructions that come with Api Life indicate that two treatments a year are recommended (once in spring and again in autumn), I have found that a single treatment protocol in autumn is adequate as long as it is combined with a variety of low-impact varroa controls—such as trapping, making splits, genetically resistant bees, and screened hive bottoms—throughout the rest of the season. Without the aid of these low-impact controls, a single autumn treatment of Api Life may not always suffice unless another highly effective treatment is applied to reduce mite levels in early spring. Unfortunately, in my experience, adding Api Life to nucleus colonies when they are freshly created in the spring seems to slow down their development. It almost seems as if the essential oil vapors cause the queen to slow down or stop laying eggs for a period of time. This occurs even though the average daytime temperature falls within the proper treatment range as described on the label. It may be that the smaller population of the newly formed colony is not as effective at keeping the hive adequately ventilated as compared to larger, well-established hives. Whatever the reason, I recommend the use of Api Life only as a fall treatment and suggest some other high-impact protocol when a spring treatment is desired.

Another essential-oil-based treatment consists of a thymol-containing gel sold under the trade name Apiguard. Like Api Life, the gel is introduced into the hive above the brood chamber. Research indicates that this product is effective against varroa in a variety of geographic locations and at different times of the year.[9] Apiguard is reported to help control tracheal mites and chalkbrood and, like other essential oil treatments, is temperature sensitive. Beekeepers have also reported some success by adding essential oils to sugar syrup and feeding this mixture to their hives. Clearly this activity should not take place when honey supers are on the hive; and the regular use of sugar syrup as a carrier for essential oils does not fit in with the organic philosophy of beekeeping, for reasons that were discussed in the section on feeding in chapter 3.

Food-Grade Mineral Oil

Although I have never ventured to try it myself, food-grade mineral oil (FGMO) has received a fair amount of attention as a treatment for varroa mites. Research indicates that the use of straight FGMO offers some benefit against mites, though not enough to successfully allow mineral oil to be the sole control method used against varroa.[10] This contradicts numerous testimonial reports of good results against varroa from beekeepers using FGMO around the United States. The basic treatment protocol incorporates an electric or propane insect fogger as the mineral-oil-dispersing mechanism.

Used by itself, it seems that FGMO has limited effectiveness in the area of mite control. In much the same way as the grease patty works, a mineral oil fog spreads a light coating of oil throughout the hive, which provides an impediment to sure footing for the mites. It also might provide enough irritation to increase levels of grooming behavior among the bees, which may in turn dislodge a greater number of varroa from their hosts than would otherwise be the case. In this regard, it seems logical that any use of FGMO should be combined with a screened bottom or screened bottom board to capitalize on its potential benefits. It is possible that a fogging of mineral oil could also coat some of the mites with oil, clogging up their breathing tubes and suffocating them in the process. Indications are that FGMO can be used most effectively as a carrier for more potent acaricides, such as thymol, that have proven toxic to mites but safe for people.

The downside to the use of FGMO applied with an insect fogger is the threat of fire posed by the fogger, especially when used in dry conditions, or during periods of drought. Due to its flash point, mineral oil has the potential to ignite and potentially injure bees should a blast of flame be emitted from the fogging mechanism. It was

> **FGMO Emulsion Recipe**
>
> Here is a mineral oil–thymol–honey emulsion formulation that is reported to have successful results.[12]
>
> *Ingredients*:
> 1 L (33.8 oz) food-grade mineral oil
> 1 kg (2.2 lb) beeswax
> 1 kg (2.2 lb) honey
> 100 g (3.53 oz) thymol
> 50 ml (1.69 oz) ethyl alcohol (minimum 90%)
> 100 1-meter-long (39.37 in.) pieces of cotton cord (or 6/32-in. upholsterers' welt cord)
>
> *Preparation:*
> 1. Heat mineral oil and then add beeswax. Stir well until wax melts completely.
> 2. Remove from heat and mix in honey until honey is dissolved.
> 3. Dissolve thymol in alcohol.
> 4. Thoroughly stir diluted thymol into oil/wax/honey mixture
> 5. Add cotton cordage pieces to mixture and allow them to become fully soaked.
> 6. Store in a sealed container in a cool place until ready to use.

also noted by researchers that potential health hazards to operators, from both inhalation and exposure to the skin, are unknown.[11]

A much safer method for utilizing mineral oil in its capacity as a carrier for stronger mite-killing substances is to soak cotton cordage in an FGMO-thymol-honey emulsion and lay it on top of the brood nest. The addition of honey to the mineral oil mixture attracts bees, who promptly engage in chewing up and removing the emulsion-soaked cotton cord. Whether fogger-applied heated FGMO, or any of its by-products, may contaminate honey stored in combs or be absorbed by wax are questions that remain to be answered. In theory, the cool application of mineral oil using cotton cords should pose less of a concern with regard to any contamination issues.

The Sugar Solution

For the honey producer, it is something of an irony that a safe, effective way to control mites has been developed that makes use of sugar. Sugar has long been a major competitor of honey when it comes to the shopper's grocery dollar. This situation is a sore point for many commercial beekeepers given that the U.S. beekeeping community receives no government subsidies for the production of honey and the all-important side benefits of crop pollination, whereas U.S. sugar producers receive subsidies that are often in excess of the world's market price for sugar.

This new, nontoxic, natural alternative to chemical insecticides is based on sugar esters, which act to both suffocate the mite and break down the outer waxy coating that covers the varroa's body and normally keeps it from shriveling up from loss of moisture and dying from dehydration. As such, these esters must come in contact with the mites to kill them. These sugar esters are unique in that their lethal effects wash quickly over the target pests and yet they leave no harmful residues in their wake. Nontoxic sugars and fatty acids are all that are left after the tide of death recedes.

The commercial result of this breakthrough technology for beekeepers is a product called Sucrocide. Just follow the directions and add the appropriate amount of this biochemical miticide to water, fill up a basic garden-variety sprayer, hose the bees down, and watch the mites fall. Although the product is relatively safe to handle, care must be taken in its application, because it is possible to kill honey bees when higher-than-recommended dosages are used. Even after proper treatment with the correct concentration and quantity, the bees will look "washed up" and bedraggled, but with time they will recuperate and return to normal. The key to preventing a bee-killing overdose is to calibrate your sprayer so that the prescribed one and a half ounces of biopesticide is applied to each deep frame. A good way to calibrate has been outlined by the Washington State University Department of

Entomology in the July 2004 *American Bee Journal*.[13] Their procedure calls for measuring six ounces of water into a disposable container and marking the level on the side. With the sprayer adjusted to deliver a fine mist, determine how many seconds it takes to fill the container up to the six-ounce mark. For example, if it takes sixteen seconds to fill the container, then the rate of discharge is one and a half ounces every four seconds. At this rate, each side of a full-depth frame should receive a two-second blast of Sucrocide. The nozzle of the sprayer should be kept at a distance of about nine to twelve inches from the frame during application. To ensure even coverage, make sure to direct the spray across the frame at a consistent and steady rate.

This treatment protocol results in soaked honey bees, and, like people, honey bees do not fare well when they become wet and chilled. Therefore, it is not recommended that Sucrocide be applied when temperatures are expected to drop below 55°F (13°C). Because the biopesticide must come in contact with the mites to be effective, treatments applied when the bees are clustering will inhibit effective application.

Sucrocide can be applied at a faster rate by simply spraying the appropriate amount down between the frames, rather than removing and spraying each frame individually as directed by the label. The Florida Department of Agriculture & Consumer Services Division of Plant Industry has developed a commercially viable advancement to this approach. It is a spray wand that incorporates a manifold with nine nozzles attached to it that allows a liquid to be applied between the frames from the top of a hive in a single pass.[14] Although not quite as effective as spraying each frame individually, this fast method requires much less labor and works well enough that, with regular applications, adequate control has been reported. Care must be taken that the amount of brace comb and burr comb in hives is kept to a minimum so that it does not impinge on the efficacy of these topside applications.

Sucrocide is composed of sucrose octanoate esters, which are synthetic sugar esters and therefore require special regulatory approval to be utilized in certified organic production. However, the high level of efficacy and incredibly low toxicity of this biopesticide should allow it to be added to the NOP's National List of allowed synthetic substances, thereby making it available for future use in organic honey production.

Organic Acids

In general, the use of acid to control varroa is ideal because it is corrosive but not toxic, and it does not bioaccumulate in honey or wax, thus avoiding contamination issues. Several acids, such as formic, acetic, oxalic, and lactic (which is utilized in human metabolism), are all found naturally in honey.[15] As a result, federal regulators exempt such acids from tolerance requirements for honey and beeswax. Because acids used for mite control act physically on the target organism, the chances of the mites building up resistance is remote. When applied appropriately to a hive, acids are beneficial against both varroa and tracheal mites. These characteristics make acid an ideal candidate for use in organic hive management.

Although not toxic, acids are extremely caustic and can burn holes in clothing or cause severe burns to exposed skin, making them dangerous for the applicator as well as the bees if not used carefully. Wearing protective clothing, such as gloves and eye protection, is strongly advised when working with acids. Care must also be taken by applicators to position themselves upwind from the acid source so fumes do not overcome them during treatments. There are four main types of acid that can be utilized as high-impact controls for varroa and tracheal mites.

Formic Acid

Formic acid is used fairly extensively in several countries throughout Europe, as well as in New Zealand and Canada, but it has only recently received approval in the United States for use against varroa mites. Not only will formic acid kill mites clinging to adult bees, but it also under certain circumstances appears to pass through cappings and kill varroa mites that are in sealed brood cells.[16] When formic is employed as a liquid that must be vaporized through natural evaporation, it is temperature sensitive. The ideal temperature range for delivery is approximately 60° to 80°F (11° to 27°C).

Beekeepers in Canada, who have typically used formic acid liquid in concentrations of 85 percent by squirting it on the bottom board and letting it evaporate up into the hive, have reported some difficulties. Apparently, due to the close correlation between dosage and temperature when using the acid, such treatments prove to be very hard on the queen, often leading to queen failure or supersedure. Most Canadians who utilize this approach now further dilute the 85 percent acid down to the 60 to 65 percent range so that the treatment will be less problematic for the bees.[17] Due to the chance that the hive may experience queen issues, in general acids are best applied in the spring, so that there is plenty of time for the hive to sucessfully supersede its queen or be given a replacement and still provision itself adequately for winter. However, a mild day in late fall or early winter has also been reported to be extremely effective for application, because the reduction in the size of the brood nest exposes the maximum percentage of mites to the treatment.

An early effort to take advantage of the benefits of formic acid while minimizing the potential hazards of application was the release of a formic acid gel pack approved for use by the FDA against *V. destructor* and distributed by Betterbee, a company based in Greenwich, New York. I had the opportunity to use these gel packs for two seasons, and they worked very well. Unfortunately, the use of the formic gel pack was short-lived, as the product was pulled from the market after the first year when it was discovered that the packaging would fall apart as a result of the action of the acid.

Since then, other methods of formic acid delivery have become available. Among the options are Mite-Away II pads. The directions that come with these pads indicate that a one-and-a-half-inch spacer rim of wood is required to provide room for the pads between the top bars and the cover of the hive, as are additional pieces of wood to hold the pads up off the top bars of the frames that make up the brood chamber. In actual practice, however, the pads seem to work fine when placed directly on the top bars of the brood chamber once the burr comb has been removed. The wooden rim spacer can also be disposed of by simply flipping over the inner cover so that the flat side of the cover faces up and the side with the rim faces down. While this additional space does not respect the colony's bee space, it does provide room for the pad. Other formic delivery systems include the MiteGone evaporation system developed by Mr. Bill Ruzicka; the Nassenheider Evaporator; plastic bags containing Homasote (tentest board), developed by the Ontario Beekeepers' Association; as well as various evaporators made by do-it-yourselfers.[18]

Oxalic Acid

Oxalic acid can be utilized in much lower concentrations than formic acid, making it a safer alternative for the naturally inclined beekeeper. Aerosol applications of oxalic have been used successfully for many years throughout Europe and have been promoted in the United States by biodynamic enthusiast Gunther Hauk of the Pfeiffer Center in upstate New York. Treatment consists of lightly spraying the bees with a fine mist of a 2.1 percent dilution of oxalic. The treatment protocol calls for the application of a half ounce per side of each brood frame. The correct concentration may be obtained by mixing one ounce of oxalic acid with a quart of water. Because the low concentration of acid will not penetrate into sealed brood cells, oxalic acid is best applied when there is a minimum of brood and the majority of varroa within the hive will be exposed to the treatment, such as during the winter months.[19]

Unlike treating with formic acid, each frame in the hive containing bees is traditionally removed and sprayed with a very light mist of oxalic that covers the bees without wetting them down. This is a very labor-intensive procedure and often requires the beekeeper to enter the brood nest at a time when the natural inclination is to leave the bees alone and not disturb the winter cluster. Gunther Hauk reports, however, that such treatments are not harmful to wintering bees and when correctly performed may result in 98 percent effectiveness, even in temperatures as low as about 40°F (4.4°C). A new oxalic acid evaporation device developed by the Swiss company Andermatt Biocontrol AG should make oxalic acid more attractive to those with concerns about the amount of labor involved in treatments and the effect of opening up the hive and breaking the cluster during winter applications.[20] U.S. beekeepers can obtain an oxalic acid vaporizer from Heilyser Technology of Sidney, British Columbia. (For more information, please visit the resource section at the end of this book).

As an alternative to spraying each frame individually, one can dribble a mixture of 3.5% concentration of oxalic acid in sugar syrup made up in a ratio of one part sugar to one part water. This can be obtained simply by mixing 35 grams of oxalic with one liter of sugar water. This solution is trickled down between the combs and onto the bees with the use of a syringe or garden sprayer. Daytime temperatures should be at least 38°F (3°C) during application which is typically conducted once each season, in autumn. This dribbling method is much faster and easier than hand spraying each frame individually and is well suited to larger commercial operations. The fact that no fine particles are formed by a spray mist and may be accidentally inhaled makes dribbling a much safer option for the person doing the application.[21]

Lactic Acid

Long appreciated for its nutritional benefits when consumed in lacto-fermented foods such as sauerkraut, lactic acid is another acid that has been used by beekeepers with good results against mites. Application of lactic acid is similar to that of oxalic acid, except a concentration of 15 to 18 percent is used. Four or five treatments are applied throughout the year. When used during the winter when temperatures are over 50°F (10°C) and no brood is present, a 99 percent mite kill can be expected.[22]

Acetic Acid

Acetic acid (vinegar) is also being used to control varroa and tracheal mites. As with some of the other organic acids, users report suppression of various hive diseases in addition to mite control. Strong vinegar, with a 20 to 25 percent dilution of acetic acid, is typically used, along with a vaporizing apparatus for application. One commercially available unit manufactured by Better Way Wax Melter Honey Processors incorporates a blower working in conjunction with the vaporizer to help distribute the acid fumes throughout the hive.

Active Biological Varroa Controls

Biological control of varroa relies on life itself to deal effectively with the destructive nature of this pest, thereby bringing a sense of natural balance back into the hive's ecosystem. Efforts to develop hygienic bees that are naturally resistant to mites have been the focus of the majority of the biological control efforts within the beekeeping community. However, other examples of biological controls are being worked on and hold great promise.

Small-Cell Foundation

Some folks have proposed the theory that small-cell foundation can be used to improve the ability of *Apis mellifera* to withstand the presence of varroa as well as other diseases and pests. It has been well established that the cell size of brood comb directly affects the size of the bee that develops within it, with larger cells producing bigger bees and smaller cells reducing the physical size of the emerging adult. Two leading proponents of the small-cell approach are Ed and Dee Lusby of Arizona Rangeland Honey. Ed and Dee are reportedly the first commercial beekeepers to successfully overwinter bees on a regular basis without the use of high-impact varroa mite treatments of any kind. They claim their current success is a result of their converting to small-cell foundation. However, it is important to note that during the process of transitioning to small-cell foundation only about 10 percent of their colonies survived the initial reduction, and it was these hives that were used to rebuild their colony numbers.

According to the Lusbys, most, if not all, of our current beekeeping woes can be traced back to the period between 1880 and 1939, when a concerted effort was made to increase the cell size of bee foundation to produce larger honey bees. The reasoning was that this would result in an increase in honey production, because bigger bees are capable of carrying larger loads of nectar and have a longer proboscises to help them reach deeper into blossoms. As a result, today's foundation is typically made to accommodate a 5.4-millimeter or larger cell size. The major problem with this artificial enlargement, according to Ed and Dee, is that although the bigger cell size has increased the average body size of honey bees and the amount of honey they are able to produce, it has decreased the vitality of the honey bees' immune system, making them generally more susceptible to disease and parasitic mites—the full consequences of which we are experiencing today.

A skeptic would argue that the reason Ed and Dee have been able to forgo the use of chemicals and antibiotics is more likely due to the fact that their hives are located in Africanized honey bee territory, and either they have picked up some of the AHB's natural genetic resistance to mites, or their European honey bee queens have actually been replaced by their Africanized cousins. The intense heat of the Arizona desert also plays a role in creating a less hospitable environment for the healthy growth and reproduction of *V. destructor*.

Despite these potential mitigating factors, there is strong evidence to support the Lusbys' claims and the theory of healthier bees as a result of smaller cell size. In the first place, the average cell size of foundation used in European honey bee colonies has indeed been artificially increased over the years. In the early days of foundation manufacture, cell size equaled approximately five cells to the inch, just shy of 5.1 millimeters per cell.[23] Amos Ives Root (1839–1923), the founder of the A. I. Root Company, hired a die maker to manufacture the first foundation rollers in the United States. Although initially the foundation mill was built on the basis of 5 cells per inch, Mr. Root later decided to increase the cell size to 4.83 cells per inch (approximately 5.26 millimeters per cell).[24] Cell size continued to be enlarged to the point where today's standard wax foundation typically has cells that are about 5.4 millimeters in size. However, cell size will tend to vary somewhat, because foundation is prone to stretching during the manufacturing process or when being installed into frames.

The second piece of evidence to support the theoretical benefits of controlling varroa with small-cell foundation stems from the fact that *V. destructor* does not cause great damage to smaller species of bees that naturally live closer to the equator, such as varroa's natural host, the Eastern (or Indian) honey bee, *Apis cerana*, or *Apis mellifera scutellata*, the infamous Africanized honey bee. The Indian and African honey bees will bite at mites and remove them from their bodies, which is one way they combat varroa—but they also have smaller worker-comb cell diameters and abbreviated developmental times for the workers to grow from eggs into adults when compared to the European honey bee. This shortened pupation period limits the number of varroa that are able to reproduce successfully and helps keep their population in check. This suggests that *A. mellifera* raised on a smaller-sized cell foundation may also have a shorter birthing cycle from egg to adult, which would provide a measure of natural mite control similar to that enjoyed by their cousins living nearer to the earth's equator.

Getting back to a cell size that is closer to *A. mellifera's* natural state, as proposed by the Lusbys, is not a simple proposition, if for no other reason than a queen bred on and accustomed to laying eggs in 5.4-millimeter cells is likely to be too large to comfortably fit her abdomen into 4.9-millimeter cells. Indeed, bees raised in and on 5.4-millimeter cells when building comb naturally, without the benefit of foundation, seem

141

to construct cells that measure about 5.4 millimeters. According to Dee Lusby, worker bees raised on large-cell foundation encounter significant difficulties when trying to draw out foundation of a smaller cell diameter than that upon which they were raised. To deal with this complication, Dee recommends making the transition in two steps. First, reduce the cell size of the comb down from 5.4 to 5.1 millimeters, and then, after a few years, once the bees have stabilized on the new comb, reduce their cell size again down to 4.9 millimeters. It is important to remember that the honey bee's current cell size is the result of incremental cell enlargement that began over one hundred years ago. We can't expect to reverse such an ingrained and established biological process overnight.

One of the numerous challenges associated with this endeavor is obtaining foundation with the appropriate cell size. Normally I am not a fan of plastic foundation, because the bees don't take to it nearly as well as they do to pure beeswax foundation. Foundation manufacturers have attempted to get around this issue by applying a coating of beeswax to the plastic. Either way, the use of plastic is prohibited by most organic regulations. However, this is the one occasion where plastic foundation or comb, with a 5.1-millimeter cell size, may be useful and appropriate in an organic operation on a limited basis while making the transition from a larger cell size to a smaller one. Alternatively, if they can be obtained, packages of bees or nucleus colonies raised on a smaller-cell comb that matches your hive's cell size can be utilized with good success.

Fungi

Not being native to Vermont and the Champlain Valley, I have noted with a degree of consternation the pessimistic tendencies of most of the beekeepers here, which tends to match that of farmers in this neck of the woods. This attitude was eloquently expressed to me by beekeeper Bill Mraz, who noted shortly after the arrival of *Varroa destructor* that if we really wanted to control the mite, we should start trying to raise and propagate the species. It would then be just a matter of time, he postulated, before some type of disease or critter would come along and start killing off the mites we were trying to raise. There may be something to this pessimistic attitude shared by so many in Vermont agriculture, because it turns out that just such on organism has been uncovered by researchers at the ARS Beneficial Insects Research Unit

(BIRU) at Weslaco, Texas, who have worked with a strain of the fungus *Metarhizium anisopliae*. This particular fungus has proven itself to be highly pathogenic to *Varroa destructor*, yet it is not toxic to humans and does not harm bees or interfere with the queen's reproduction. Although the fungus does not work immediately and takes three to four days after being introduced to a hive before peak mite mortality occurs, the acaricidal effects have been found to continue for at least forty-five days following the application of the spores. The fungus was observed to spread to other colonies through the natural drifting of workers and drones between hives, a situation that is likely to prove beneficial.

Although the fungus does not seem to penetrate the cappings on sealed brood cells, treatments are not limited to periods when no brood is being produced or when brood levels are low, given that its extended time of activity within the hive allows it to be effective over the course of several of the mites' reproductive cycles. According to the researchers, *M. anisopliae* was as effective as the miticide fluvalinate (the active ingredient in Apistan) when used on mites that were not resistant to fluvalinate.[25]

When it comes to high-impact mite treatments, the ideal is a substance that is deadly to varroa, is effective over long periods, is not toxic to bees or humans, and will not contaminate honey. *M. anisopliae* seems to fit this description well. In addition, the use of biological controls that do not carry the potential for the buildup of resistance by the mite harmonizes with the organic philosophy of honey bee husbandry. As a result, *M. anisopliae* is a potentially exciting tool that the organic beekeeper may have available in the future, at least until the day when a truly varroa-resistant honey bee becomes available to make such measures unnecessary. The major challenges that must be overcome if *M. anisopliae* is to become a useful tool in honey bee management is to produce spores that have an adequate shelf life, as well as to develop an effective way of introducing the fungi into the hive so that it comes into contact with enough mites to make it a viable control option.[26]

Tracheal Mites

The honey bee tracheal mite (*Acarapis woodi*) was first identified in England in 1921, and its discovery spurred the creation of the

Honeybee Act of 1922 by the United States Congress. This legislation regulated the importation of new bee stock into the United States in an effort to prevent the introduction of tracheal mites, as well as other pests and diseases. It performed the job admirably well until 1984, when mites infesting the breathing tubes of honey bees were found for the first time in the United States along the Mexican border.

The U.S. beekeeping industry, led by Roger Morse of Cornell University, focused in the late 1980s and early 1990s on breeding strains of bees that showed levels of resistance to tracheal mites. As a result of these types of efforts, *A. woodi* is not considered a major threat by most beekeepers today. With luck, our tracheal mite experience may prove to be a foreshadowing of what we can expect from *V. destructor*. As with its larger cousin, *A. woodi* was considered a minor problem in its home country but caused large-scale losses to the American beekeeping industry during the first dozen years or so following its introduction on American shores.

Detecting *A. woodi* infestation is complicated by the microscopic size of the mite. Tracheal mites are visible only with the aid of a microscope or optic lens. As a result, a quick, easy, and widely utilized method of mite identification in the field has yet to be developed. One indication of widespread tracheal mite presence is numerous bees within a hive that have formed their wings into what appears to be the letter *K*. Honey bees have two pairs of wings, and each pair is hooked together when they fly. Bees can disconnect their wings for grooming purposes, but the telltale *K* orientation is usually found only when mites are in attendance. This sympton is thought to be associated with a virus that the mite carries and spreads among the bees.

Although not a definitive diagnosis, the observation of bees crawling around outside the entrance of an unmolested hive when temperatures are low or below freezing is another sign that a heavy infestation of *A. woodi* is probably present. The bees act as if their tracheae are so clogged up with mites that they are having trouble breathing, and they go outside to have better access to fresh air. Bees suffering from this condition will often linger around the outside of the hive's entrance so long that the cold will work upon their joints, making them stiff and immobile, until they eventually lose their footing and flop over into the snow, where their numbers pile up throughout the course of the winter. Thus, the all-too-typical sign of tracheal mite infestation is a large number of dead bees lying on the ground in front of the hive early in the spring after the snow melts. However, it is possible that other conditions may cause bees to behave similarly, so the only sure way to test for the presence of *A. woodi* is to capture some live bees and inspect them under a microscope. Check with your state bee inspector, as he or she may provide this service for free under certain conditions.

Workers spotted hanging around the hive entrance when temperatures are below 40°F (4.4°C) may be an indication that tracheal mites are present.

As with all pests and diseases of the hive, natural tolerance or resistance is the best form of control. Buckfast bees from England are widely held to contain genetic tolerance to *A. woodi*. Aside from breeding, beekeepers have long utilized the fumes given off by menthol, a component of peppermint oil, for the treatment of hives infested with tracheal mites. The only commercially available menthol treatment for tracheal mites is Mite-A-Thol. This product is made up of menthol crystals that cause the microscopic mite to die from dehydration. However, because the menthol fumes must reach inside the trachea of the honey bee, Mite-A-Thol can be used effectively only when average daytime temperatures are warm enough to allow the menthol to evaporate (between 60° and 80°F/16° and 27°C).

The grease patty is a homegrown remedy for tracheal mite infestations (see recipe on page 128). Because of *A. woodi*'s microscopic size, the oils from a grease patty, when lightly covering the bodies of the bees, inhibit the movement of the mite from bee to bee, which out of necessity must occur outside of the honey bee's breathing tubes. As a result, grease patties work to help prevent the spread of mites from infected bees and limit the damage they may cause within a colony. Due to the ability of an oily surface to hinder the movement of the tracheal mite, it is likely that a fog of food-grade mineral oil sometimes used for varroa control could prove to be an effective agent in the control of *A. woodi*.

145

Notes

1. H. Shimanuki, Nick Calderone, and D. Knox, "Parasitic Mite Syndrome I. The Symptoms," *American Bee Journal* (1994): 827–28.

2. Joe M. Graham, ed., *The Hive and the Honey Bee* (Hamilton, IL: Dadant & Sons, 1993): 1121.

3. A. I. Root Co., *The ABC & XYZ of Bee Culture* (Medina, OH: A. I. Root Co., 1990): 322.

4. Ibid.

5. U.S. Environmental Protection Agency, "A Summary of the Emissions Characterizations and Noncancer Respiratory Effects of Wood Smoke," EPA-453/R-93-036, as cited in Citizens for Safe Water Around Badger, "Fact Sheet: Open Burning at Ravenna Arsenal," http://www.cswab.org/ravenna.html.

6. Jerry Hayes, "Sugar Dusting," *American Bee Journal* 145 (January 2005): 25.

7. S. Hart, "Baby Bee Odor Lures Cradle-robbing Mites," *Science News* (August 12, 1989): 103.

8. Ibid.

9. R. Rivera, F. A. Eischen, R. H. Graham, and G. M. Acuna, "Varroa Control Trials with the Thymol-based Gel Product," *American Bee Journal* 145 (May 2005): 433.

10. P. J. Elzen, R. L. Cox, and W. A. Jones, "Evaluation of Food Grade Mineral Oil Treatment for Varroa Mite Control," *American Bee Journal* 144 (December 2004): 921–23.

11. Ibid.

12. Pedro P. Rodriguez and C. E. Harris, "Food Grade Mineral Oil— Thymol Widen Alternatives for Honey Bee Parasite Control," *American Bee Journal* 143 (September 2003): 727–30.

13. Department of Entomology, Washington State University, Dandant advertisement in *American Bee Journal* 144 (July 2004): 526

14. Reed E. Burns, Justin E. Emerson, and Gerald B. McElroy, "An Economical and Effective Spray Device for Applications of Liquids to Beehives," *American Bee Journal* 145 (October 2005): 803–804.

15. Joe M. Graham, ed., *The Hive and the Honey Bee* (Hamilton, IL: Dadant & Sons, 1993): 877–78.

16. Malcolm T. Sanford, "Using Liquid Formic Acid for Mite Control," *Bee Culture* 131 (June 2003): 17–19.

17. Ibid.

18. Ibid.

19. Gunther Hauk, *Toward Saving the Honey Bee* (San Francisco, CA: BIO-Dynamic Farming and Gardening Association, Inc., 2002): 70–71.

20. Ibid.

21. Dr. Eva Rademacher and Marika Harz, "Effectiveness of Oxalic Acid for Controlling the Varroa Mite," *Ameican Bee Journal* 146 (July 2006) : 614–617.

22. Gunther Hauk, *Toward Saving the Honey Bee* (San Francisco, CA: BIO-Dynamic Farming and Gardening Association, Inc., 2002): 70–71.

23. C. C. Miller, "Stray Staws," *Gleanings in Bee Culture* 35 (September 1907): 1127.

24. E. R. Root, "80 Years Among the Bees, The Evolution of Modern Comb Foundation," *Gleanings in Bee Culture* (May 1948): 295.

25. Lambert Kanga, Rosalind R. James, and Walker Jones, "Field Trials Using the Fungal Pathogen, *Metarhizium anisopliae* (Deutermycetes: *Hyphomycete*) to Control the Ectoparasitic Mite, *Varroa destructor* (Acari: *Varroidae*) in Honey Bee Colonies," *Journal of Economic Entomology* 96(4): 1091–99.

26. Rosalind R. James, Gerald Hayes, and Jarrod E. Leland, "Field Trials on the Microbial Control of Varroa with the Fungus *Metarhizium anisopliae*," *American Bee Journal* 146 (November 2006): 968–72.

CHAPTER 6

INSECT PESTS

His labor is a Chant—
His idleness—a Tune—
Oh, for a Bee's experience
Of Clovers, and of Noon!

—Emily Dickinson

Insects make up about 80 percent of the animal population on Earth. Unfortunately, beekeepers have a tendency to place their primary focus in the apiary on the honey bee and remain unaware of the other insects in the vicinity until the hive is under attack. Then we rush in to exterminate the offenders without realizing that we might also be harming other beneficial bugs, including the bees themselves.

As organic beekeepers, we should attempt to learn as much as possible about the behavior and biology of the other insects that live in and around our apiaries. Frequently, a little knowledge about how certain insects function can provide us with the ammunition we need to prevent pest damage before it becomes necessary to use invasive and potentially toxic control methods.

Wax Moths

A strong aversion to maggots is not a good trait to have if you intend to keep honey bees. A couple of insects that will give you plenty of opportunity to practice getting over your maggot repugnance are the greater wax moth (GWM), scientifically known as *Galleria mellonella*, and the lesser wax moth (LWM), *Achroia grisella*. Prior to the arrival of the small hive beetle, the wax moth was America's greatest threat to unprotected honeycombs during periods of prolonged warm weather. The adult wax moths will lay their eggs among the combs, or in cracks and crevices that the bees residing in the hive are unable to reach. Once they hatch, the wax moth larvae will feed on the combs and their contents—cast-off honey bee larva skins, pollen, and honey. As a result, they prefer older dark combs that contain high levels of impurities and larval castings rather than newly created combs that feature the beautiful golden yellow coloring of freshly formed beeswax.

The moth larvae bear a resemblance to the maggots typically found on carrion, reaching a maximum body length of about one inch, though their final size is largely dependent on the type and amount of food available. As they grow, the feeding larvae will chew their way through the combs, leaving silk-lined tunnels of webbing to mark their path. This webbing becomes littered with the wax moth larvae's feces and gets so thick that it shields them from any nest-cleaning worker bees that may try to sting or remove the larvae from the hive. Once the larvae mature, they spin a cocoon and pupate in a protective site within the hive. The cocoons are usually spun in oval depressions that the larvae chew into the surface of the wood, and into which they adhere themselves. The adult wax moth will then emerge from the cocoon to begin the cycle all over again. The speed of growth of the larvae and the pupae depend a lot on the temperature, with the process taking much longer when cool temperatures prevail than during warm periods.

Mother Nature's cleanup crew, the wax worms, can make short work of combs that are uninhabited by a sufficient number of honey bees. Photo by Steve Parise.

In Vermont, the danger of GWM and LWM damage begins primarily in July, after the moths have either migrated from the South or hitched a ride north with migratory beekeepers. The threat lasts well into September and even October, until the first hard frosts kill off the adults, eggs, and larvae. The wax moth can survive the cold northern winter only by living on combs in a heated building. In the South, though, the wax moth is a threat on a year-round basis. Combs that have become home to wax moth larvae can become totally ruined if not caught in time. If found early enough, most of the webbing may be torn off the comb and the frame placed into a strong hive that will remove the remaining webbing, as well as clean up, repair, and replace damaged comb if the destruction is not too severe.

It was while pulling a chunk of webbing off an infested frame to expose the maggotlike larvae squirming around in the center of the comb for the first time that I felt a great revulsion and—I must admit—a wave of nausea that rippled through my stomach from the unappetizing sight. Although I do not wish wax moth infestation on anyone's hives, I am happy to report that one's initial revulsion subsides with repeated exposure to these little critters.

To control the damaging effects of the wax moth without chemicals, the organically minded beekeeper can make use of two things that keep most everything healthy: sunlight and fresh air. Adult moths prefer to lay their eggs in dark crevices deep within the hive, and they are reluctant to lay in combs that are exposed to light and fresh air. Because of this, protection from the moths may be obtained by simply storing empty hive bodies and supers of combs on end with the tops and the bottoms of the frames exposed to the sunlight and ventilation.

By far the best form of control for GWM and LWM are the bees themselves. A strong hive will do a fine job of removing any moths or larvae that are unfortunate enough to try to make a home in their hive. It is the weak hive that is susceptible. A common mistake among novice beekeepers when they find a dead colony filled with moth larvae is to think that wax moths killed their hive. The wax moth is an opportunist, a scavenger to whom nature has delegated the task of cleaning up dead hives, thereby acting as a natural form of disease prevention to help keep other colonies in the neighborhood healthy. The only way wax moths can be found thriving in a hive occupied by bees is if the honey bee colony lacks sufficient numbers to adequately defend and patrol the entire cavity composing the hive's interior. For this reason, I believe it is a good practice to manipulate the size of the hive to match its occupants. I make it a practice to add empty supers to a hive only when a colony has almost filled the space it has. This way the area that the bees must keep free of invaders is minimized. The exception to this practice comes after the honey harvest, at which time I place the freshly extracted supers, still wet with honey, between the inner cover and the outer cover of the hives. The hives are typically very strong and populous at this time of year, and they are often overcrowded given that they now have to occupy a smaller space than that which existed prior to the removal of the excess honey supers. Such hives have more than enough bees to crawl up through the inner cover into the empty supers above, cleaning up leftover honey and prohibiting wax moths from gaining a foothold in the stack of empty combs.

> The wax moth is an opportunist, a scavenger to whom nature has delegated the task of cleaning up dead hives, thereby acting as a natural form of disease prevention to help keep other colonies in the neighborhood healthy.

The one major drawback to this approach, however, is that if there is a late honey flow, a strong colony will often start to store the freshly gathered nectar in the supers above the inner cover. These honey-filled combs are more likely to hold uncapped, unripe honey, as the cool evenings tend to prevent the bees from maintaining their honey-ripening duties around the clock. As a result, such honey, if harvested, must typically be used within a month or so—otherwise it may start to ferment and pick up a vinegary taste. This can sometimes be avoided if the bees combine this high-moisture honey with a significant amount of the remaining fully ripened honey that is left over on the wet frames that have gone through the extraction process and have been stored above the inner cover. Depending on how efficient your extracting process is, this leftover honey may be enough to offset the high moisture content of the unripe honey, preventing fermentation.

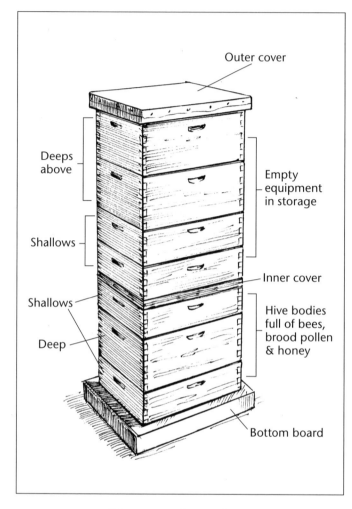

Outer cover

Deeps
above

Shallows

Shallows

Deep

Empty
equipment
in storage

Inner cover

Hive bodies
full of bees,
brood pollen
& honey

Bottom board

Keep wax moths out of your extracted equipment and allow the bees to clean up the leftover honey on the combs by storing them on the hive until the first hard frosts arrive.

Because all my honey-extracting equipment has been cleaned up and put away by the time the first hard frosts arrive, supers that become filled with late-season honey are typically placed on a hive that contains borderline honey supplies to give them additional honey stores to get through winter. However, it is not a good idea to use a partially filled super for supplemental feeding at this time of year because the bees are unlikely to have the chance to fill out the remaining frames with honey. The winter cluster may then eventually move into the area that contains the empty comb and starve.

If none of your hives can make use of an additional super of honey, late-season honey supers can be placed a couple hundred feet away from the apiary on an unseasonably warm day, allowing the bees to rob out the supers prior to your placing them inside for winter storage. Leaving the supers containing honey any closer to the apiary is likely to create a feeding frenzy that will encourage bees from neighboring hives to steal honey from one another. This is often referred to as *robbing*.

I have also found that when storing extracted supers on top of hives, it is preferable to place shallow supers directly above the inner cover and position deep supers higher up on top of the shallows. This is because queen bees seem to prefer to lay their eggs in the larger span of comb that is found within the frames of the hive bodies, and deep supers stored directly on top of inner covers following the harvest often end up with brood in them by the time they are ready to be put away for winter. In this situation the entire hive body, if filled up 75 percent or more with brood and honey, can be placed below the inner cover and left as part of the overwintering hive. Again, under no circumstances should empty or partially filled honey supers (less than 75 to 80 percent full) be left on top of a hive for the winter season. To do so would invite the cluster to move up into them and risk starvation late in the winter or early spring.

Another form of wax moth control can be found in the form of a special strain of the microscopic creature known as *Bacillus thuringiensis* (Bt). Available commercially in a product manufactured by Vita called B401 (Certan), Bt is 100 percent effective in controlling wax worms in empty supers of drawn comb. B401 must be used as a preventive measure, however; it will not be as effective after an infestation has taken place. A single application of a diluted Bt concentration sprayed over the wax surface of each comb will provide protection for up to a year or more, until the equipment is placed back into use. Because *Bacillus thuringiensis* becomes toxic only when exposed to the unique digestive environment found within the wax worm's gut, it is a safe alternative to mothballs (paradichlorobenzene), which can contaminate honey and wax.

> To control the damaging effects of the wax moth without chemicals, the organically minded beekeeper can make use of two things that keep most everything healthy: sunlight and fresh air.

Although the U.S. Environmental Protection Agency (EPA) has approved paradichlorobenzene for use to protect honeycombs from wax moth damage, as a synthetic chemical that has not been included in the NOP's National List, it is not allowed in organic production. However, wax moths should not be a major problem for the conscientious beekeeper who takes the time to perform regular inspections of hives and takes steps to protect empty supers that are not occupied by bees. Keeping all equipment filled with healthy bees is the best way to confront the wax moth threat. In the Champlain Valley of Vermont, this means that all supers of drawn comb should be in use and inhabited with bees by about Independence Day. If conditions prevent frames of drawn comb from being used early enough during the course of the season, it is a good precaution to spray each frame with a Certan solution, fumigate them with carbon dioxide, store the combs where they will be exposed to fresh air and sunlight, or place all unused supers in a freezer for at least twenty-four hours prior to storing the equipment in such a way that the wax moths cannot gain access to them. These are the only methods of control that are acceptable to organic certifiers. Should you find yourself faced with equipment full of baby wax moths despite your best efforts, I hear the worms make great fishing bait.

Small Hive Beetles

The small hive beetle (SHB), technically known as *Aethina tumida*, is indigenous to Africa. It was first discovered in the United States at the end of the twentieth century. *A. tumida* beetles are reddish to yellowish brown in color immediately upon emergence from their pupal stage. Once fully mature, the adult beetles turn dark brown to nearly black. They are about one-quarter inch or so in length, slightly smaller than a ladybug, although beetles of varying sizes have been reported and are probably the result of differing food sources and climates. *A. tumida* is distinguishable from most other similar beetles in that the midsection of its body containing the wings is sectioned off from both the head and the tail areas in such a way that the insect looks like it is composed of three distinct segments. Its small size, quickness, hardness, and covering of fine hair make the beetle difficult to pick up by hand.

SHB are similar to wax moths in that they are opportunists that prosper by scavenging on weak or dead hives, as well as stored equipment that contains drawn comb. Thus, it is not wise to leave stacks of empty equipment sitting around. Sanitary conditions around the bee yard should be maintained by removing broken combs, burr comb scrapings, and so on, because they may act to attract beetles and support their reproductive cycle. This advice also applies to the honey house. Supers filled with honey should not sit too long before being extracted, and it's not a good idea to let cappings sit exposed for extended periods.

Beetle eggs are about 1.4 millimeters long, 0.26 millimeters wide, and similar in appearance to honey bee eggs. Adult beetles will lay irregular masses of eggs in cracks and crevices throughout a hive or stack of equipment. The eggs typically hatch within one to six days depending on the relative humidity and temperature. SHB eggs will not hatch below 50°F (10°C) and require humidity levels above 50 percent for larvae to successfully emerge.[1]

154

Once hatched, the SHB larvae feed on honey, pollen, eggs, and brood until they mature and then crawl out of the hive and tunnel into the ground to pupate. Beetle larvae do not weave the same messy mass of webbing that is characteristic of wax moths, and upon first glance they resemble the common maggot even more than wax worms do. They also move faster than wax moth larvae. Luckily, the SHB larvae do not consume and destroy a lot of beeswax like the wax moths. The larvae tunnel through the combs in search of food and can heavily damage newly drawn-out and delicate comb; however, older, sturdier brood combs are better able to withstand heavy larval infestation without disintegrating. The greatest damage from the beetle comes from the destruction of the honey stored within the combs as the immature beetles eat their way through the hive and the converted nectar becomes contaminated by the larval feces and ferments in the comb. Infested honey bee colonies have been observed to abandon combs and even entire hives once they have become badly infested and the fermentation process has begun.

Small hive beetles have been known to overwinter within the honey bee cluster in northern latitudes. Photo by Jeffrey Lotz.

A major factor affecting the spread of SHB are temperatures that resemble the beetle's tropical-subtropical native climate in Africa. The beetle is considered to be a significant problem in the warm southern states such as Georgia and Florida, but it remains to be seen how much of an issue it will be in other regions, where cooler temperatures prevail. Like all insects, SHB have a minimum temperature below which reproduction and development will not occur. In addition, the beetle's growth and metabolic rate will increase as the temperature increases.

SHB's population density within a given geographic location is even more reliant on their success at breeding and overwintering, which in turn is tied to the number of degree-days that the temperature range is above the minimum needed for reproduction. Consistent reproductive success of the SHB requires temperatures above 68°F (20°C), the temperature at which approximately 50 percent of the developing beetle larvae will survive. At temperatures below 50°F (10°C), small hive beetle larvae were found not to survive at all.[2] This is likely to be the major factor that prevents SHB from becoming a major issue in northern regions.

Additional factors that play a role in healthy beetle growth and development are soil moisture levels and density. SHB require soil moisture levels of 5 to 25 percent to pupate successfully.[3] This would suggest that areas in western states that experience prolonged periods of dry weather, areas that see a lot of rain over lengthy periods, or areas where the soil type is such that moisture from local precipitation will accumulate over time will not support beetle reproduction and therefore are unlikely to encounter significant long-term problems from SHB. Choosing yard sites that receive direct sunshine will play a role in helping to keep moisture levels in the soil low. Soil compactness is also significant, simply because loose soils are much easier for SHB pupae to burrow into compared to dense hardpan or clay soils. It is likely that populations of beetles that survive from year to year and have the potential to grow in numbers are most likely to occur in locations that experience prolonged warm temperatures and have soils that are composed primarily of silt, which will hold some moisture and does not get extremely hard, dry, or waterlogged.

In contrast to the wax moth, which must use stealth to enter the hive, lay its eggs, and then depart before being discovered and killed, *A. tumida* has the ability to move about inside the hive and even coexist to some degree with the bees within a colony. Honey bees of European stock have long been bred for gentleness, making them much less aggressive than their African relatives, and it is likely that this difference in behavior accounts for some of the contrast in severity of SHB damage between Africa and North America.

However, the increased propagation of hygienically inclined honey bees may be helping to reverse this trend, as the aggressively hygienic bees mimic the imprisonment behavior that is a typical SHB defense expressed by the African Cape honey bee, *Apis mellifera capensis*. Guard bees will corner adult beetles in confinement areas built out of wax and propolis and prevent their escape. The imprisoned beetles are thus prevented from mating, laying eggs, and doing further feeding damage to the hive. Small hive beetles will starve within about two weeks without food, and yet imprisoned beetles have been known to survive up to two months without access to food in combs.[4] This is

156

because *A. tumida* has the ability to touch a worker bee's antennae in such a manner as to fool it into feeding the hungry beetle. As a result, small hive beetles and worker bees can often be observed scurrying around on the top bars of hives immediately after the inner cover is removed on a hive that is infested. The act of popping the inner cover frees the beetles that had been held captive between the cover and the top bars. Frequent hive manipulations in areas that support large SHB populations may lead to higher beetle activity and additional stress on the hive due to the repeated release of the beetles from captivity.

If not for the complication of beekeeper interference, European honey bees that exhibit a strong tendency to collect propolis might prove to be more adept at confinement behavior and be better adapted to tolerate beetle infestation. In the event that beetle populations grow to high levels within a colony, European honey bees will again imitate African Cape honey bees through their tendency to abscond, leaving their old hive behind and finding a new home.

The first efforts of which I became aware to control SHB were homemade traps that consisted of a small hole drilled into the bottom board and an attached jar beneath the hole filled with beer, vinegar, or mineral oil to attract the unwelcome hive guests. The beetles, in their search for a corner of the hive to hide in, would enter the jar through the hole in the bottom board and drown. Another device that has been partially successful in controlling the small hive beetle is a hanging trap baited with fruit such as cantaloupe and similar in design to traps used for yellow jackets. The success of this approach is limited, however, due to the fact that comb filled with brood, honey, and pollen is much more attractive to beetles than fruit, which serves as their secondary feeding source. Unfortunately, honeycomb as bait within a trap has not proven much better than fruit at attracting adult beetles, because the comb bait in the trap is no more attractive to the beetle than the comb in nearby hives.

The first commercially available trap that came onto the market specifically for use against the hive beetle was the West Beetle Trap. This trap consists of a tray filled with vegetable oil and covered with a grid that contains holes large enough for the beetles to pass, but too small for the bees. The trap is placed on the bottom board of the hive; wooden spacers are needed to raise the hive body above it about two centimeters and provide clearance for the bees to enter and exit while the trap is in place. It is also critical that the base of the hive be leveled prior to setting the trap in place and adding the vegetable oil, to prevent the oil from leaking out and possibly killing the bees. Dadant & Sons of Hamilton, Illinois, recommend checking the trap about every two weeks and performing periodic cleanings of the device. A significant benefit of the trap is that repeated manipulations of the colony that disturb beetles in hiding or release them from imprisonment cause them to move around the hive and potentially find the trap sooner. A side benefit of the West Beetle Trap is that any varroa mites that lose their footing and fall from above are likely to land in the oil and perish along with the beetles.

Dr. Mike Hood of Clemson University in South Carolina developed another commercially available trap that builds on the original jar-under-the-bottom-board idea. The Hood Small Hive Beetle Trap attaches to an empty frame that has not been fitted with foundation. The trap has separate compartments that are filled with apple cider vinegar or mineral oil and replaces one of the frames positioned next to the side of the hive body containing the brood chamber during the summertime, or in an upper super during the winter. Beetles that enter the trap are attracted by the vinegar, drown in the oil, and are easily removed.

Aside from having to open up the hive in order to check and clean out the trap, the major disadvantage of the Hood Small Hive Beetle Trap is that by replacing a frame within the hive, it creates less room for the bees to store honey or raise brood. Also, you should be prepared to spend time cleaning the extra comb the bees will build in the open area of the frame surrounding the trap, as well as any propolis the bees may use to try to plug up the entrances to the beetle-killing device.

Another form of *A. tumida* control that some folks are recommending is removal of beetles by hand, either by crushing them with a hive tool or by sucking them up with a small vacuum. Although anecdotal reports from hobby beekeepers claim excellent results using such manual techniques, their efficiency is questionable, and the amount of time such control measures would take when relied upon exclusively makes these methods unrealistic for commercial beekeepers. However, this inefficiency does not in any way diminish the strong sense of immediate gratification and accomplishment that comes from personally defending the bees by crushing small hive beetles with a hive tool as they scurry around looking for cover when a colony is first opened.

The use of pheromones has been suggested as a beetle control mechanism because male beetle pheromones are effective in controlling other beetle species. Also, male SHB have been reported to enter honey bee colonies before the females, which indicates the possibility that similar hormones are involved in colony invasions.

Another strategy that has received a lot of attention in the search for adequate SHB control is targeting the larvae when they enter their roving stage in search of adequate ground to pupate in. Some of the more natural management techniques that focus on the larvae include the use of moats, setting larva traps around the hives, sprinkling the ground around infested colonies with diatomaceous earth, and allowing chickens to roam around the hives. Biological controls include the use of soil fungi (*Aspergillus*) and nematodes that will attack small hive beetle larvae while they are pupating in the soil. Beekeepers who frequently move their hives to new apiary locations have reported reduced beetle problems, presumably because they break the beetle's reproduction cycle and eliminate ready access to hives for beetles emerging from the soil following pupation. Unfortunately these approaches all seek to address infestation after the fact, and none of them prevents larval damage to the hive or equipment.

Luckily there has been success in renovating combs that have been slimed with larvae feces and drip with fermented honey. Measures that seek to salvage equipment after infestation by SHB larvae include household products that kill beetle larvae and improve honey bee acceptance of affected combs. These products include regular bleach, white vinegar, dishwashing detergent, and vegetable oil. A dilution of 50 percent bleach is reportedly 100 percent effective in killing larvae within four hours, while a mixture of dish detergent at 1 percent strength kills 85 percent of the beetle larvae within twenty-four hours.[5]

I do not spend a lot of time and resources attempting to trap or kill the small hive beetle adults or larvae. However, I do admit to sometimes crushing a few with my hive tool as they try to scurry away immediately after being exposed when the inner cover is popped open on an infested hive. For the most part, my SHB control management has been limited to simply not leaving any empty equipment lying around the bee yard once the weather turns consistently warm. I am also careful to extract harvested honey supers within several days of taking them off the hive, and so far this strategy has worked out well. I find that spotting the occasional hive beetle crawling around on frames of honey that the bees and I have worked so hard to produce and harvest provides more than enough motivation for me to extract the honey as soon as possible and banish any thoughts of procrastination.

One fall when I was in the process of packing the hives up for winter, I came across a colony that had died out sometime after I had harvested its excess honey and applied an essential-oil mite treatment approximately three to four weeks earlier. Due to robbing, there was not a lot of honey left in the frames, and when I entered the brood area within the hive and pulled out a frame for inspection (to be sure the colony had not died from disease), I was confronted with comb that was swarming with maggotlike SHB larvae. This was my first encounter with beetle larvae. Many years of encountering wax moth larvae had served me well, as I felt no disgust or nausea at the sight. In fact, I found it fascinating that, when looking down between the frames from above, the combs looked perfectly normal, with no sign of damage or anything out of the ordinary. It wasn't until I had removed and tilted the frame and looked into the comb that the hundreds of developing beetle larvae became visible in their womb, deep within the cells along the midrib of the waxen structure. As the faint smell of fermented honey wafted up, I quickly placed the frame back into the hive body from which it came.

> As a result of the small hive beetle, the days of stacking hives high with empty supers early in the season and returning at the end of the season to harvest them are over. . . . [H]ives should receive an empty super only when they have the majority of their current space filled.

Now I felt panic setting in. I had never dealt with combs full of these creepy, crawly critters before. Luckily, I had not yet read the numerous warnings that I would later come across that indicate stacking supers containing beetle larvae onto healthy hives is a sure way to spread the infestation from one colony to another. Nor was I aware of the "authorities" who suggest that bees are reluctant to clean up combs filled with fermented honey and prefer to abandon such combs instead. Since I was not quite sure what to do, I treated the dead hive filled with SHB larvae as I would have treated a hive infested with wax moth larvae, in which the frames of comb would still be salvageable. I placed the boxes of larvae-filled frames on top of the inner covers of the strongest hives in the bee yard and hoped for the best.

160

Upon returning to the yard a week later, I had visions of maggots having overrun my strongest colonies, oozing out of every orifice of the hive as the former colonies were slowly consumed like the dead, bloated carcass of an animal that has not been discovered by the turkey vultures and other scavengers. Much to my relief, however, all signs of the beetle larvae had disappeared like magic, and although the frames of wax comb invariably had suffered some damage, on the whole they were still intact and appeared undamaged, as if my whole experience the previous week had been nothing more than a bad dream, with only the components of the dead hive sitting on top of the colonies providing proof of the reality of my experience. A quick check of the empty combs revealed no sign of SHB larvae or of any fermented honey, and all of the hives upon which the infested supers were placed overwintered well and were in fine shape come spring.

If the dire warnings not to stack or store infested equipment on top of strong colonies is to be believed, then I got lucky. Perhaps my efforts to incorporate hygienic queens into my apiaries and raise bees with hygienic behavior have been more successful than I realize. (I have never actually tested any of my hives specifically for the hygienic trait, relying instead on the test of winter survival.) Then again, maybe the cool September temperatures hampered the larvae's growth and vitality, making them more vulnerable to the defending bees. I imagine that because most of the honey in the supers had been robbed out before the beetles moved in, the bees had an easier time restoring the slimed combs to usable condition than would have otherwise been the case. Whatever the factors involved, I am glad I chose to break up the dead hive and place only a single infested box on top of each strong colony. Stacking all the supers containing larvae on a single hive could have been detrimental, because such a large excess of combs might well have proven too much for the bees to properly defend and clean up.

As a result of the small hive beetle, the days of stacking hives high with empty supers early in the season and returning at the end of the season to harvest them is over. Even before the days of varroa, this was not a good hive management program anyway, because the bees would typically fill out the center frames of each super and ignore the combs on the outer edges of the colony. Of course, for those who didn't have the time to occasionally visit their colonies during the course of the season, this approach was a necessity. With the arrival of the SHB, however, hives should receive an empty super only when they have the majority of their current space filled. The rule of thumb I use is to try to time the addition of a new super to a colony to coincide with the bees starting to work on the outermost frames on either side of the uppermost box after filling and capping most everything else below and between. In this way, the bees have enough space for additional honey storage and to keep them from feeling overcrowded, but not so much extra room that they are unable to prevent parasites from taking up residence.

Spiders and Other Insects

Various spiders and other insects may live in, on, or around beehives, and many will feed on bees, pollen, wax, honey, brood, hive debris found on the bottom board, or even the wooden hive parts themselves. Fortunately most pests are only local or occasional in nature, and most do not require special control measures.

Many insects such as cockroaches and earwigs can be found living in beehives but do no apparent harm. However, these insects can be very offensive if they make their way into a jar of extracted honey. Some spiders and larger insects such as dragonflies and praying mantises will capture foraging honey bees, but they seldom cause significant losses to a colony. The same can be said for bullfrogs. Spiderwebs on or around hives are often littered with the remains of dead bees and can be easily taken down with a flick of a hive tool.

Various types of ants can often be found living in or around the hive. Most of these ants live off hive debris and are more of a nuisance than any kind of serious threat. Termites and carpenter ants that attack wooden hive parts can be a severe problem, however. An organically appropriate deterrent to these pests is an oil barrier. This requires a hive stand positioned so that the legs of the stand sit in containers of oil; another option is to apply a grease coating to each leg of the stand to create a barrier to the pests. Such arrangements have also proven effective in protecting hives from deadly fire ants, which have been known to invade and destroy bee colonies in southern climes.

Hive stands have the additional benefit of keeping colonies up off the cold, wet ground, thus helping to keep hives warmer. The onset of wood rot is also delayed by reduced contact with the damp earth. Stands that raise the hive entrance above the level of grass and other vegetation help ensure that the honey bee's flight path is not interrupted in the event that the beekeeper gets behind in mowing around the beehives. In general, keeping the grass mowed and the bee yard clean of debris will go far in reducing the likelihood that ants and other critters will take up residence.

Notes

1. Doug Somerville, NSW Agriculture, Rural Industrial Research and Development Corporation, Publication No. 03/050, RIRDC Project No. DAN-213A, "Study of the Small Hive Beetle in the USA" (June 2003): 5–7; and M. Michael Hood, "The Small Hive Beetle, *Aethina Tumida*: A Review," *Bee World* 85(3): 56.

2. Somerville, "Study of the Small Hive Beetle in the USA," 6.

3. Ibid.

4. Hood, "Small Hive Beetle," 54.

5. Ibid., 57.

CHAPTER 7

FOUR-LEGGED AND FEATHERED PESTS

*My grandfather always said that living is like
licking honey off a thorn.*

—Louis Adamic

Many of the creatures in the animal world that wield offensive weapons—poison glands or powerful claws and teeth—are colored conspicuously to warn potential adversaries to leave them alone. The bee's striped colors are effective in advertising such a message, so much so that other insects such as the drone fly and bumblebee moth, both stingless creatures, manage to evade the attention of many predators because they possess coloration similar to that of bees. When looks are not enough to deter a would-be predator, the honey bee has the capacity to resort to a cooperative team effort in an attempt to discourage a marauder. Despite the considerably intimidating defense a beehive can muster, a number of animals have developed defenses and strategies that allow them to feast on a hive's contents.

Skunks and Raccoons and Bears, Oh My!

Although not normally considered a major threat to the honey bee, skunks (*Mephitis mephitis*) can weaken strong hives and cause lethal damage to already weakened colonies. The skunk is renowned for scratching around the hive entrance, swatting the bees that come out to see what all the fuss is about, and then contentedly munching on them in their dazed and injured state. Skunks are active primarily at dusk and dawn from late autumn through early spring; signs that skunks have been dining on your tab include scratch marks around hive entrances, matted grass and scratched-up earth in front of hives, skunk tracks, and small piles of feces in and around the bee yard that resemble cat droppings but contain bee parts. However, the very first signal that skunks have made your bee yard a regular stop on their rounds is often the dozen or so bees that start buzzing around your head with a tone of angered annoyance as soon as you show up. Colonies that are molested by skunks, or anything else, on a regular basis will tend to be very jumpy and behave in a highly irritable manner.

Colonies that are molested by skunks, or anything else, on a regular basis will tend to be very jumpy and behave in a highly irritable manner.

Although much less prevalent, raccoons (*Procyon lotor*) have been known to cause damage in the apiary. Though known to feed on insects, the raccoon is more likely to reach up under stacks of empty supers stored in an apiary and rip apart the frames of comb for a sweet snack. The raccoon closely resembles the bear but is not large enough to knock over hives and rip apart supers filled with bees and honey. Damaged frames of drawn comb removed from stacks of empty equipment and scattered around the bee yard are signs that raccoons have been at work.

One very effective way to discourage skunks and raccoons is to take some pieces of plywood and fill one side with nails or screws that stick out a quarter to a half inch on the opposite side and are spaced about a half inch apart. The plywood should be at least eighteen by twelve inches in diameter. These boards can be placed in front of each hive to make it uncomfortable for any critter that wants to pull up a seat and snack on your bees or equipment. There have been instances where such boards have been pushed aside by a determined animal. To prevent this, it's a good idea to embed a few sticks or tent stakes in the ground on either side of the boards to help anchor them and to prevent their easy removal. Just remember to remove the boards or turn them over when working around the hives, or you run the risk of nails penetrating your shoes.

A similar approach that removes the inconvenience of having to move the skunk boards when working with the bees is the use of carpet grippers or tack strips, which are available at your local carpet store. These thin wooden strips have tacks sticking out of one side so that, when they are nailed to the floor along the walls of a room, wall-to-wall carpeting is stretched over the tack strips and tucked into the edge between the strip and the wall to hold the edge of the carpeting in place. When cut to the width of a bottom board opening and nailed in place on the area of the bottom board that extends out in front of the hive, the carpet gripper becomes an effective skunk deterrent. When the carpet tack strip is nailed to the alighting area with the beveled edge facing away from the hive, the angle of the tacks will point outward and will be most effective in discouraging skunk feeding activities. In areas prone to frequent skunk activity, making these strips a standard part of your bottom board construction will go a long way toward easing your mind of concerns over skunks.

Stake down your skunk boards, or the crafty critters are likely to move them out of their way.

Another method that works well is to place some metal fencing material in front of the hive entrance. Sheep fencing rolled into cages for tomatoes works great when performing double-duty for this purpose. A loose coil of fencing creates enough of a barrier that critters can't easily forage at the hive entrance.

Some folks place their hives up on stands that raise the colony about three feet off the ground to prevent access to the hives by skunks. A hungry skunk that must stand on its hind legs and expose its vulnerable underbelly to feed on a raised hive can quickly become

166

discouraged. Keeping your hives on a stand is also likely to reduce the amount of bending that is necessary to work the hives, but a small ladder or step stool may be required during a good honey year.

No wild animal causes more fear and trepidation among beekeepers than the bear. Even the smallest of these in the United States, the black bear (*Ursus americanus*), can wreak havoc and major destruction within an apiary. Due to its size and strength, the bear is capable of the greatest destruction and economic damage because it not only destroys bees and combs but can easily crush and tear apart frames, hive bodies, and supers.

As with other threats to the honey bee, prevention is the preferred method of controlling a marauding bear. Once a bear has gotten a taste of the gourmet delicacies packaged inside those cute little wooden boxes sitting within easy reach, about the only way to stop it from coming back again and again until the whole bee yard is devastated is to shoot the bear, capture and remove it to another location far away, or move the bees. Clearly the latter two options represent the more humane and sustainable approach. However, the substantial amount of time and resources involved with such endeavors can make the inconvenience of erecting some type of bear deterrent around the bee yard seem well worth the effort. If bears have been sighted around your area, or you observe tracks, feces, or scarred trees that have been used as scratching posts, do not delay in providing some bear-proofing to your apiary. Otherwise the next signs of bear activity you are likely to come across are flattened beehives that have been torn apart and eaten.

The value of simple preventive measures cannot be overstated. First and foremost, as when dealing with skunks and raccoons, a clean bee yard will be less inviting to a curious bear. Leaving old supers

A skunk would have to be pretty handy with a hammer to remove a tack strip nailed to a bottom board.

Large stones or cinder blocks placed inside the coil of wire will prevent these barriers from being moved by a hungry skunk.

167

or frames lying about the area will make your bees more noticeable and therefore a more likely target. Bears typically prefer wooded areas and are often found in mountainous regions; by simply locating bees away from prime bear habitat whenever possible, much damage can be avoided. If you must place bees in bear territory, bear-proofing measures are a must. One of the first things that should be done is to register the bee yard with your state bee inspector. Many states and Canadian provinces will provide some compensation to beekeepers who lose their bees to bears, as long as the bees have been properly registered.

> Instances abound where bears have sacked hives kept in populated areas close to homes, even where dogs are in residence. Hence, the first order of defense requires that the bear be strongly discouraged from getting that first fateful taste of honey.

While I have so far been fortunate enough to have never experienced a foraging bear in my bee yards, I have heard enough horror stories to know that it is unlikely that there is any such thing as a 100 percent bear-proof fence once a bear has gotten a taste of the contents of a bee yard. Reports claim that a determined bear will tear down, walk through, climb over, or burrow under anything that stands between it and the golden ambrosia contained within the lemon yellow wax combs of the hive. And an apiary does not have to be located in an isolated wooded area in order to become a target. Instances abound where bears have sacked hives kept in populated areas close to homes, even where dogs are in residence. Hence, the first order of defense requires that the bear be strongly discouraged from getting that first fateful taste of honey.

There are a number of options available for protecting one's bees from bears. A common choice is the electric fence. This is primarily because an electric fence is typically easier to install than a permanent fence, and it is easier to move in the event that the bees must be transported to another location. If the bees are expected to permanently reside in a single location, a permanent fence can be used. No matter what type of fence you choose, rugged construction is essential due to the bear's powerful nature. Wooden or steel posts should be buried deep in the ground. It is also advisable to use extra-thick fencing wire and ceramic, rather than plastic, insulators on an electric fence. Place the insulators on the inside corners of the posts so they will be less likely to break or be pulled out of the posts should an angry bear swat its paw at the wiring. Four or more strands spaced at approximately one-foot intervals and a fully charged battery round out your bear protection. To help ensure adequate grounding, some folks recommend that every other wire be grounded. If the bee yard is near enough to a power supply, the fence can be plugged right in. Solar panels are a great way to help ensure that the battery stays fully charged in isolated locations. Be sure to keep the grass and weeds well trimmed under the wires, or a short circuit may develop that will compromise the effectiveness of the fence. A side benefit of an electric fence is that if the lower strand of wire is placed low enough to the ground, it may also be effective in deterring skunks and other critters. Another idea that Bill Mraz suggests is that strands of barbed wire be used on electrical fencing in place of the standard wiring so as to further deter an animal from pushing through the fence.

I have heard folks recommend hanging a strip of bacon, or some other bait, from a hot wire so that a bear approaching the fence will touch the tempting morsel with its nose or tongue (both wet and sensitive organs) and get a good zap that should discourage further investigation. However, it seems to me that hanging bait on the fence is likely to make the bee yard much more attractive to a curious bear than when no bait is used at all.

A low-tech approach involves creating a bed of sharp thorns that will deter a hungry bear in much the same way as was recommended for skunks and raccoons. Pieces of plywood containing nails or screws long enough to protrude between one-half and one inch and embedded on one side of the board at two- to three-inch intervals create a powerful, low-cost, low-maintenance deterrent when compared to a fence. The boards should be at least two feet square or larger and placed all around the hive so that a curious bear will not miss them when approaching from any direction. Care should also be taken to prevent the crafty bear from easily catching the edge of a board with its paw and turning it over or shoving it out of the way. Pushing the bear boards up flush against the hive or adjacent boards and staking them with wooden sticks or tent stakes around the edges can accomplish this. I have also heard of some beekeepers using carpet remnants in place of plywood. Although carpeting has the advantage of not warping like wood does, care must be taken to use thick enough carpeting to hold the nails or screws with enough rigidity that they will not be easily pushed over.

As with the skunk boards, bear boards or carpeting will need to be moved or turned over when you are working around the hives. Using large enough sheets of plywood that are not easily shoved out of the way or flipped over works well as long as they are not too difficult for you to move when entering the bee yard.

Mice

Mice are not quite as much of a problem down South where winters are not so long and cold, but in the northern reaches of the United States, a nice, relatively warm and dry beehive filled with wonderful things to eat proves a strong temptation for mice. If given the chance, mice will move into the bottom super or the hive body of a colony after the temperatures are consistently below 50°F (10°C) and the bees have formed their winter cluster, leaving the majority of the hive's interior uninhabited and unprotected. The mouse can build its nest on the cluster's periphery with little chance of being molested until the temperature increases up above the 45° to 50°F (7° to 10°C) range.

This in itself wouldn't be so bad if the mice would refrain from chewing holes in the wax combs and damaging wooden frame parts while building a nest in the hive. Along with nest debris and holes in the combs come little mouse turds and urine. As long as the hive is strong and healthy, the mice will keep their distance and stay in the bottom portion of the hive, well below the cluster above, since the bees don't like having mice in their midst any more than most beekeepers do. Should the weather warm up to the point where the bees can break their cluster and venture down to investigate the area that the mice have annexed, the mice will simply leave until the temperature falls again and it is safe for them to return. However, should the hive find itself in a weakened condition or die out completely, the hive you expect to be bustling with bees in the spring will be transformed instead into a little mouse condo.

Reducing the width of all hive entrances that are large enough to admit a mouse, so that bees can still pass but mice are excluded, will prevent such disappointing springtime discoveries. A piece of half-inch hardware cloth three inches wide and cut to the width of the bottom entrance provides excellent mouse protection when bent at a forty-five-degree angle lengthwise and wedged into the space between the bottom board and the first bee box (see picture on page 167). Mouse excluders manufactured from pieces of wood with holes through which the bees can come and go are also available from bee supply companies. Unfortunately, they will also reduce airflow, negatively affecting ventilation within the hive during the crucial winter months.

The key to the efficacy of any mouse guard, however, is positioning the device in the entrance *prior* to the onset of cold weather. The tardy beekeeper who installs mouse guards late in the season after the first cold weather has arrived may find that, instead of keeping the mice out, the mouse guard has served to lock the mice inside, guaranteeing that the damage to the hive will be worse than it would have been without a guard in place at all. As a result, it is best to install mouse guards on warm, sunny days when the bees are flying strongly. This way any mouse that may have moved in during a recent cold snap will have deserted its nest due to the increased bee activity and will be waiting for the bees to reform the cluster in the cool of the evening before returning to its anticipated winter abode.

> The tardy beekeeper who installs mouse guards late in the season after the first cold weather has arrived may find that, instead of keeping the mice out, the mouse guard has served to lock the mice inside, guaranteeing that the damage to the hive will be worse than it would have been without a guard in place at all.

You may benefit if there are rock walls, wooded areas, and other places that make good habitat for snakes near your apiaries, as many snakes will happily make a meal out of any mouse they come across that they can fit in their mouth. However, in places like Florida, extra care must be taken because many of the snakes that live in the area are venomous and like to hide underneath the beehives. They can provide quite a surprise to anyone who happens to be moving hives around and carelessly reaches underneath a hive without first making sure the area is unoccupied.

If you are employing the service of snakes to help keep the mouse population in check around your bee yards, keep a sharp eye out when mowing the grass within the apiary. One day while mowing a bee yard, I spooked a garter snake that had been hiding amongst the hives. I watched as it ducked into the bottom entrance of a nearby beehive, only to exit less than a minute later looking like a porcupine, bristling with freshly laid stingers throbbing and pulsing all along the length of its back. The poor snake was not moving anywhere near as fast upon exiting as it had when it entered the hive, and it became more and more sluggish, dying a short time later.

Don't let all the action at the hive entrance distract you from what may be happening behind the front lines in this battle against the mice either. Those empty honey supers, full of frames that the bees worked into drawn comb and that you have extracted, can be very attractive accommodations for a mouse. After all, there are no bees in these empty stored supers to worry about. Given that organic standards frown on the use of poisons, physical barriers are your best defense here. Organic honey standards require a pest-free honey house and extracting area. Prevention is the first line of defense, utilizing barriers and good housekeeping. Control comes later if preventive measures are unsuccessful.

Prevention starts with a solid building. Inside, consider stacking supers kept in storage on top of one another, with the bottom super sitting on a flat surface, and use something mouseproof like an outer cover to seal up the top of the pile. Be sure to keep the stack straight, with each super lined up directly on top of the one below. It takes only one askew box in the pile to create a large enough opening for a mouse to squeeze through. If preventive measures fail or additional peace of mind is required, the assistance of some traps or the employment of a cat or two can be useful, though both these approaches tend to work slowly and may allow for some damage to occur before all the mice have been removed from the area. Although it may not seem to be in keeping with organic philosophy, the use of poison is often permissible in organic production as long as it can be demonstrated that the rodenticide will not affect the integrity of the final product in any way.

The Birds and the Bees

For the most part, birds are a minor difficulty for beekeepers outside of Africa and Asia, which are home to avian species such as bee-eaters, the beeswax-eating birds called honeyguides, and at least two species of bird that will lead honey bee predators such as honey badgers or baboons to a hive and then dine on the spoils left behind after the larger mammal has exposed the nest and eaten its fill. In North America, heavy local predation of honey bees may occur by species of birds such as titmice, swifts, flycatchers, and some shrikes, which may present a problem for queen breeders and others working to rear their own mother bees.[1]

I have also observed on several occasions, while working in one particular bee yard located just a couple miles from the Canadian border, a flock of seagulls circling high above the apiary for long periods, and I have wondered whether they were dining on some high-flying bees. Other than the possibility of seagull slaughter or the capture of a queen on her mating flight, the bird that catches the occasional bee on the wing in North America does not present any serious threat to beehives, with one notable exception: the woodpecker.

Although not a common occurrence, a woodpecker will occasionally peck holes in a hive to gain access to the tasty honey bees within. A woodpecker can seriously harm a weak hive, though the primary damage these birds cause is to the hive equipment itself. The one instance of this that I observed was damage that occurred after the fact when a woodpecker had opened up a large hole in and around the thin layer of wood located by the handhold on a hive body. The size of the hole was fairly large, which seems to indicate that the woodpecker spent a considerable amount of time feeding at the hive, either in one sitting or during repeated visits. The damage was discovered early in the spring and no further woodpecker activity was observed during the course of the season, which suggests that perhaps the woodpecker attacked the hive during the cold weather, when other insects were scarce and the number of honey bees that would emerge from the hive to investigate such a disturbance were few and slow-moving. Presumably, once the number of bees venturing forth to look into the situation increased with the temperature to the point where the bees were able to land a number of stings, the woodpecker lost interest and moved on to easier pickings.

Controlling damage in such an instance is difficult at best, and although I know of nobody who has experience dealing with an active woodpecker attack, it is possible that introducing a barrier, such as a cardboard box slipped over the hive, may create enough of an inconvenience to deter such activity once a woodpecker is actively engaged in feeding on a hive. Alternatively, if a power source is available nearby, a motion detector connected to a floodlight or a radio may provide enough deterrence to discourage all but the most persistent bird. When necessary, woodpecker holes can be patched up by nailing or screwing a piece of sheet metal over the inside of the opening.

Notes

1. Joe M. Graham, ed., *The Hive and the Honey Bee* (Hamilton, IL: Dadant & Sons, 1993): 1136.

CHAPTER 8

ENVIRONMENTAL
AND HUMAN THREATS

Botany Bees
Full many a tomato plant
Would never blush nor bear,
Without the bee to gallivant
And shift some pollen there.

She travels in the honey line,
But sets the vines aglow;
Which shows the finest things we do
Are not the things we know.

I do not care for honey much,
And yet I prize the bee;
The fair tomatoes that I love,
She makes 'em blush for me.

—Chicago Daily News

Challenges to a honey bee colony's existence are not limited to just other animals. Additional threats can appear in the form of vegetation and human activity that can sap the vitality of the hive and lead to its demise. Organic beekeepers should be aware of the types of plants within foraging range of their apiaries and the human activities that will occur nearby. Such threats may go beyond endangering the health of the hive—they may also damage the organic integrity of the products that will be harvested from the apiary. Therefore, the apiculturist must maintain a level of awareness that extends far beyond the world of the bee yard and reaches out into the surrounding region and human community.

Mold and Mildew

Mold and mildew are much more reviled by humans than by the honey bee, and they can often be found on the combs in dead hives or on the outer fringes of colonies that are weak, especially in early spring. Such a discovery is unsightly and will often cause concern among beekeepers new to the craft who are worried about the spread of disease. There is no need for concern, however, because mold and mildew are not contagious and will not cause a hive to die out. Mold and mildew grow within the hive only when the humidity is too high. In such cases it may be found on wooden interior parts of the hive, wax combs, or dead bees or brood still in the comb. It is important to remember that, like the wax moth and small hive beetle, mold moves into the hive following the decline of the bees and is not the cause of the colony's woes. However, mold may make it difficult to identify the cause of a hive's demise, as it may mask the signs of diseased brood and brood cells.

Despite their distasteful appearance and odor, hive bodies, supers, and frames of drawn comb that contain mold or mildew do not need to be discarded, nor do they need to be cleaned other than to remove burr comb that may have been built upon the insides of the supers or on the top, bottom, or side bars of the frames. It is also helpful to dislodge clumps of dead bees that may still be clinging loosely to the face of the comb in a hive that has died out before returning the now-empty equipment into service. The mold will not damage the wax, and moldy frames placed within a colony will be cleaned and polished up by the bees in no time. In fact, a strong hive can rehabilitate moldy old combs and put them back into service so fast it will make your head spin.

Poison Plants

By and large the plant kingdom offers few threats to the honey bee. The greatest may well be the threat of having a large branch from a nearby tree fall on top of a hive during a storm. Plants do produce a myriad of toxic compounds, however, and these can spill over into the nectar and make their way into honey on occasion. In 1936, Burnside and Vansell identified a dozen instances where honey bee mortality was reported from sources of yellow jessamine (*Gelsemium sempervirens*), California buckeye (*Aesculus californica*), locoweed (*Astragalus*), California false hellebore (*Veratrum californicum*), and leatherwood, also known as titi (*Cyrilla racemiflora*).[1] Reports of poisonous honey obtained from rhododendrons date back to antiquity, and they appear more recently in E. F. Phillips's book on beekeeping published in 1928, which also names mountain laurel (*Kalmia latifolia*) as a source of toxic honey. However, a large number of honey bees are kept in the Appalachian Mountains, where many of these plants are found, and there are no consistent reports of serious problems. According to Phillips, "If all the honey from these sources were poisonous there would be an epidemic of poisoning annually in this region."[2] This is not the case, though, and it is likely that the dilution of toxic nectar with nontoxic sources is a factor that reduces poisoning episodes in both bees and humans. Thus it is possible to reduce the effect of toxic nectar on a hive by feeding it honey syrup or sugar water, in an effort to reduce the concentration of poisonous honey stored in the combs. It is also reported that, in the case of rhododendron, yellow jessamine, and mountain laurel honey, only the unripe, uncapped honey is poisonous to humans; the capped honey that has been fully ripened no longer presents a danger.[3]

Most dangers from plants are inconsequential to the honey bee. It is really the beekeeper who runs the greatest risk, not from poison honey, but from plants that are poisonous to human touch and may grow in or near bee yards. In the eastern United States and Canada, the poisonous plants that the apiculturist is likely to encounter are poison ivy (*Rhus toxicodendron*) and its cousin, poison sumac (*R. vernix*), with poison oak (*R. diversiloba*) of primary concern in the western regions of these countries. Slightly less common, but no less dangerous, are giant hogweed (*Heracleum mantegazzianum*) and wild parsnip, also known

177

Poison parsnip is most easily identified in its second year of growth when it grows two to five feet tall and blooms, with flat-topped, umbrella-like, yellowish flower clusters.

Poison ivy typically grows close to the ground and features three leaves grouped on a small stem that is attached to a larger main stem. Flowers tend to be yellow or green, and the plant will bear white fruit (berries).

Poison sumac is normally found in boggy areas and can grow up to fifteen feet tall. Stems will hold from seven to thirteen smooth-edged leaves and feature glossy, pale yellow to white berry like fruits.

Poison oak can grow as a low shrub or long vine with oak-shaped leaves that develop typically in groups of three, and produces clusters of yellow berries.

Giant hogweed can grow up to fifteen feet tall and features white flowers that resemble overgrown umbrella-like blooms similar to those of Queen Anne's lace.

as poison parsnip (*Pastinaca sativa*). It is important for the apiculturist to be able to identify these plants and to locate bee yards away from areas where such vegetation is found. In the event that poisonous plants such as these are unavoidable or are discovered within the confines of established yards, there are a few nontoxic, organically approved options available to combat them. They include digging them up by the roots (with the aid of gloves and protective clothing), which is practical only if the number of plants is small; smothering them with a material such as mulch or black plastic, which works well in areas where growth is concentrated within a relatively small and accessible area; or spraying with an organically approved weed killer that won't contaminate the groundwater or soil.

Chemical Pollution

The most well-publicized danger to the hive stemming from human activity is chemical poisoning. This typically occurs from spray drift, or honey bee exposure to poisons on blossoms following an application of pesticides—this despite the fact that all pesticide applicators are required by law to adhere to label directions, which are designed to reduce the chance of honey bee death. Chemical poisoning may be identified by the dead adult bees that pile up in front of the hive entrance, often with their proboscises, or tongues, sticking out. High exposure may reduce the population of a colony to the point where there are not enough workers to care for the brood and keep the brood combs covered and warm. Pesticide spraying that must be carried out when bees are in the vicinity should be conducted at night. If spraying must occur during daylight hours, all colonies should be moved out of the area to a minimum distance of about two miles away.

A Homemade Herbicide

Here's a natural, nontoxic homemade weed killer recipe that can provide a measure of control for poison ivy. Mix one gallon of 20 percent vinegar with half a cup of orange oil and a tablespoon of natural Castile liquid soap. Spray this mixture onto the leaves of the plant. Poison ivy so treated will typically wilt and then resprout, requiring additional applications. This recipe works best during periods of dry weather, when rain will not wash the herbicide off the leaves immediately following application or nourish the roots of the plant and encourage rapid new growth.

Cautions: This contact herbicide will kill most any plant upon which it is sprayed, not just poison ivy. Be sure to thoroughly clean your hands and all tools, clothing, goggles, and protective equipment following application to remove noxious oils or acetic acid contamination. Also please note that 20 percent vinegar is much stronger than the 5 percent acetic acid content of the table vinegar we splash on our salads. It will cause skin burns upon contact, and it may cause severe lung irritation if the fumes are inhaled. Eye contact can result in severe burns and permanent corneal damage, so please take appropriate precautions to protect yourself, children, and animals.

Whereas exposure to agricultural chemicals tends to have an immediate effect on a colony, the release of everyday poisons that slowly pollute the air, ground, and water has the effect of weakening the overall immunity of the bees as well, making them more vulnerable to diseases and viruses. In this day and age, almost every species of plant and animal on Earth is suffering some form of stress from the release of pollutants into the environment. This pollution is having not only a direct effect on soil, water, and air quality, but also an indirect effect through the shifting of atmospheric weather patterns and the abnormal warming or cooling of various areas of the earth. The honey bee is no different.

> Chemical poisoning may be identified by the dead adult bees that pile up in front of the hive entrance, often with their proboscises, or tongues, sticking out. High exposure may reduce the population of a colony to the point where there are not enough workers to care for the brood and keep the brood combs covered and warm.

Biological Pollution

A new form of agriculture-related pollution with which honey bees have had to contend in the past fifteen to twenty years or so is the release of genetically modified organisms (GMOs) into the environment. Unlike traditional breeding methods that produce hybrids from similar plants and animals, genetic engineering mixes the genes from dissimilar organisms, even going so far as to mix animal, virus, and plant genes all into a single organism. Such a crossing of species would never take place naturally, and this new technology has resulted in environmental, philosophical, and moral concerns being raised that were not publicly debated on a wide scale in the United Sates prior to the release of GMOs into the environment and the food chain. Many years from now, I suspect that students of history will look back with astonishment and disbelief at the idea that humankind once had the arrogance to play around with the very foundation of all life on Earth: DNA (deoxyribonucleic acid), the carrier of genetic code. Our inability to use the technology of genetic engineering appropriately to protect non-GMO crops from contamination by GMOs threatens the very viability of organic agriculture. This conflict pits farmer against farmer—whereas the multinational agribusinesses reap the profits and have, so far, been able to avoid any responsibility for the damages inflicted by their patented technologies on non-GMO crops, as well as on livestock and people who consume GMO foods.

The engineering of the *Bacillus thuringiensis* (Bt) toxin into crops such as corn, soy, canola, and cotton is touted as an example of the beneficial use of genetic engineering. Bt has long been a mainstay of the organic farming community due to the ability of this microorganism—once consumed by the target pest and exposed to the unique environment within the digestive tract—to produce spores and crystals that cause paralysis and death. Although traditional Bt products do not kill pests as quickly as most chemical pesticides, there are no harmful side effects. Because the toxin is not produced until pests eat a leaf upon which the microbe rests, it is safe and nontoxic to animals, humans, and beneficial insects, and it may be used right up until harvest. Unfortunately, genetic engineers have installed the gene for the Bt toxin into every cell of plants such as corn and soybeans, rather than the microorganism itself. As a result, not only is the European corn borer exposed to a lethal dose of the toxin, for example, but so is everything else that feeds on the GE corn plants, from farm animals to insects and even humans, who must now adjust to consuming this poison along with their daily meals since it is integrated into the structure of the food and can't be washed off like traditional pesticide. In addition, the constant exposure to Bt has quickly led to the development of resistance among target species. This is a situation that never would have occurred if Bt had continued to be used as it has been by the organic farming community for decades prior to the release of Bt-impregnated GMOs.

Whereas exposure to agricultural chemicals tends to have an immediate effect on a colony, the release of everyday poisons that slowly pollute the air, ground, and water has the effect of weakening the overall immunity of the bees as well, making them more vulnerable to diseases and viruses.

Bees rely on plants to provide them with nectar, resins, and pollen. Pollen, being the male fertilizing seed of flowering plants and a concentrated source of genetic material, brings with it the greatest potential for GMO contamination within the hive. It is through the collection and consumption of pollen that *Apis mellifera* may suffer direct detrimental effects from genetically engineered plants, depending on the type and amount of genetically modified material it is exposed to.[4] In areas where honey bees and genetically engineered plants are in close proximity, colonies are likely to consume genetically modified material and store engineered pollen within the hive. Although not a lot of research has been done to date, some studies have been conducted on honey bees that were fed solutions of proteins expressed by GMOs. Depending on their exposure, bees had trouble learning to distinguish between the smells of flowers and exhibited a shorter-than-normal life span.[5] Such learning difficulties could also potentially make it difficult for foragers to successfully navigate back to their hive. Such belated discoveries regarding GMOs are especially troubling because once these newly manufactured genetic traits are released into the environment, they migrate, reproduce, and mutate, which means there is no way to clean up this form of biological pollution. Due to their efficiency as pollinators, the unsuspecting bees play a major role in the spread of genetically engineered material.

> Pollen, being the male fertilizing seed of flowering plants and a concentrated source of genetic material, brings with it the greatest potential for GMO contamination within the hive. . . . Due to their efficiency as pollinators, the unsuspecting bees play a major role in the spread of genetically engineered material.

Unfortunately, much of the research surrounding GMOs is filled with secrecy and controversy. This is because most of the GMO studies have been conducted by the biotech companies that manufacture and profit from these technologies, and their research is considered proprietary information and not subject to public scrutiny. In the United States, many of the government regulators charged with overseeing the approval of newly developed genetic engineering technologies are former employees of the very companies they are charged with regulating. This creates serious conflict-of-interest issues and is indicative of the charges some have leveled of lax oversight by government regulators. For example, when we look at the studies that have been done on the effects of genetically modified plants on honey bees, adult bees have been the focus of the majority of the research, yet the EPA examined only short-term honey bee larval toxicity before approving Bt-cotton plants for the U.S. market.[6]

Knowledge of what crops are being grown in the area surrounding your apiary is the only way to avoid finding yourself with hives and honey bee products that are contaminated by the products of genetic engineering. It is for this reason that most organic regulations require an apiary site map that includes the surrounding area and notes all potential threats to organic integrity.

Theft and Vandalism

Unfortunately, the risk of theft and vandalism is a real danger in today's society. One beautiful early spring day, while checking hives to be sure they had enough honey left to get them through the season, I pulled into a secluded bee yard to discover that a hunter had used a hive for target practice while passing through the area during the previous fall's hunting season. Positioning a bee yard within sight of a home rather than in an isolated, out-of-the-way place that is out of view is perhaps the best way the apiculturist can discourage criminal or mischievous behavior. Additionally, branding or otherwise marking your hive bodies, supers, and frames can aid greatly in establishing ownership of hives and equipment that have been stolen. For greater piece of mind, high-tech options offered by companies such as Bee Alert Technology of Missoula, Montana, include such gadgets as radio frequency transmitters that, when embedded in a hive, can identify a specific hive from distances of up to fifteen hundred feet with special equipment. Other security systems are able to sense when a hive is being moved and will automatically place a call to a prespecified phone number to alert the owner to any potential hive-rustling situations.

A novel approach to deterring the theft of frames of honey was administered by beekeeper David Fretz of Rutland, Vermont. After experiencing the disappearance of frames from his hives, David took a lumber crayon and wrote in bold letters on the top of each colony "CAUTION: Cyanide Experiment—Poison Honey." David reports that thefts from his apiary stopped abruptly after that.

Bees that have an extremely calm and gentle temperament may be a pleasure to work with, but it is the testy hive that is most likely to act on its own behalf to thwart attempts at thievery. Entering a bee yard and finding a hive with its outer cover removed and the inner cover askew is a sure sign that the bees have stood their ground and fought off a would-be thief. I must admit to feeling a small sense of pleasure in this kind of natural justice when envisioning the humorous scene that must have occurred to cause the intruder, perhaps bent on stealing a frame or two of honey, to forgo covering up the evidence of his or her visit and make such a hasty retreat.

Globalism

As many foreign countries are encouraged to accumulate more debt than they can ever hope to repay, in the process becoming hopelessly indebted to the United States and thus subject to undue influence on the part of the U.S. government, world trade has opened up global markets and made foreign goods more accessible to businesses. The beekeeping industry has benefited greatly from the increased genetic variety of honey bees that is now available to help develop resistant stock. One area of concern, however, is the lack of effective regulatory control surrounding the increase in global trade. It's bad enough that the market economy undercuts all social values in deference to business growth and profits, but, in addition, the wholesale exchange of goods between countries has directly contributed to the difficult situation that U.S. beekeepers find themselves in today. This is due to the increase in the number of organisms that are benign in their native lands but are allowed to immigrate, either legally or illegally, to new locales, where they become a problem.

> Tracheal mites, varroa mites, small hive beetles, and, to a certain extent, Africanized honey bees are European honey bee pests that all either were imported or migrated into the United States between 1984 and 1997.

Tracheal mites, varroa mites, small hive beetles, and, to an extent, Africanized honey bees are European honey bee pests that all either were imported or migrated into the United States between 1984 and 1997. Government regulators are appointed the daunting task of adequately controlling the movement of invasive and destructive exotic species through international trade channels, and they clearly cannot be solely relied upon. Of course, we have to remember that honey bees themselves are not native to North America. The fact that bees have had to adapt themselves to ecosystems in North America that were originally foreign to them may make them more vulnerable to certain diseases and pests within the United States and Canada.

Beekeeper Error

Apiculturists themselves may pose a threat to honey bee colonies. Beekeeper error, whether from lack of knowledge and inexperience or simply from an accident caused by not paying close enough attention, is probably the single largest cause of colony demise, greater than all other human activities combined. From not leaving enough honey on the hive to enable the bees to get through the winter to not treating a hive appropriately for diseases or mites or accidentally injuring the queen while inspecting the brood nest, the beekeeper can sabotage a hive in an enormous number of ways.

The apiculturist should make every effort to learn as much as possible from a variety of sources in an effort to reduce the possibility of mistakes. Aside from reading books on the subject of beekeeping, time spent talking with those who have greater experience is often time well spent. Local and state beekeeping organizations play a crucial role in providing opportunities for inexperienced beekeepers to ask questions of those who have been there before them. There's a lot of truth to the old saw that five different beekeepers will often have six different ideas on how best to handle a given situation. This is where your evaluation and intuitive skills come in handy, to help sift through the often conflicting information and choose the path that seems best suited to you and your particular situation. Many bee associations offer mentorship programs where a novice is paired up with a more experienced apiculturist to show him or her the ropes. Additionally, courses and workshops on beekeeping are offered throughout the country and are often a great way to obtain hands-on experience.

It is important for beginners not to allow the fear of harming the bees keep them from working with the hives. As experienced bee-keepers know, simply taking your time, paying attention, and staying focused are the best defenses against accidents and oversights. It is also important to keep up with the latest developments in beekeeping by attending bee association meetings and regularly reading bee journals where the newest research findings, industry trends, and commercial products are promoted.

Unfortunately, human error has a way of foiling the best-laid plans. By staying open to all potential sources of information and planning ahead to leave room for the unexpected and the occasional snafu, you can remove the fear of failure and recover faster from an incident. Adopting the attitude that "there are no mistakes, only learning experiences" is as helpful when working with bees as it is in navigating the trials and tribulations of everyday life.

Notes

1. Joe M. Graham, ed., *The Hive and the Honey Bee* (Hamilton, IL: Dadant & Sons, 1993): 890.

2. A. I. Root Co., *The ABC & XYZ of Bee Culture* (Medina, OH: A. I. Root Co., 1990): 263.

3. Graham, *The Hive and the Honey Bee,* 1197.

4. Louise A. Malone and Minh-Hà Pham-Delègue, "Effects of Transgene Products on Honey Bees (*Apis mellifera*) and Bumblebees (*Bombus* sp.)," *Apidologie* 32 (2001): 1–18, http://www.hortresearch .co.nz/files/science/gmimpacts/m1403malone.pdf.

5. Minh-Hà Pham-Delègue et al., "Long-term Effects of Soybean Protease Inhibitors on Digestive Enzymes, Survival and Learning Abilities of Honey Bees," *Entomologia Experimentalis et Applicata* 95 (2000): 21–29.

6. United States Environmental Protection Agency, *Bt Plant-Pesticides Biopesticides Registration Action Document,* 2000, http://www.epa .gov/scipoly/sap/2000/october/brad2.scienceassessment.pdf.

CHAPTER 9

HIVE DISEASES

If every one would do his best in watching for disease, existence would be trebly blest for all the honeybees. If every bee crank would inspect his brood-combs twice a year and with his weather eye detect the first germ to appear; then swat the thing right in the neck and knock it galley west, we'd run them off the map, by heck, if each would do his best. Bacillus larvae would not eat the baby bees alive. We'd drive them out, so help us Pete, from each and every hive. No sunken cappings would we find; no toothpick roping test; no gluepot smell, the luscious kind, if each would do his best.

—Bill Mellvir

Honey bee diseases are caused by living organisms, including fungi, bacteria, and viruses. Bees are naturally able to resist these disease organisms when they are vigorous and well nourished. The bees also make use of propolis, the most potent naturally occurring antibiotic there is, by shellacking the interior surfaces of the colony with a thin layer of the material to help maintain a sterile environment within the hive. However, despite the bees' best efforts, some microorganisms have managed to evolve to the point where they are able to establish themselves and thrive within the hive. The organic beekeeper can greatly

Healthy hives will contain brood nests with developing bees clustered closely together, all of a similar age group, and without many empty cells scattered among the brood. Photo by Steve Parise.

assist the bees in keeping disease organisms in check with practices that encourage the honey bee's natural hygienic tendencies, ensure an adequate supply of high-quality nourishment, prevent the introduction of contaminated equipment, and include regular hive inspections to catch issues early on before they develop into larger problems. Although there are a large number of diseases that can befall *Apis mellifera*, we will concern ourselves here only with those commonly encountered in the Northeast, since these are the ones I am most familiar with.

American Foulbrood

Before the advent of parasitic mites in North America, the most serious of all diseases, from U.S. beekeepers' perspective, was American foulbrood (AFB). Although varroa has overshadowed it in recent years, foulbrood is still the most important microbial disease that beekeepers must concern themselves with. As its name implies, the bacterium *Paenibacillus larvae* (formerly known as *Bacillus larvae*) that causes AFB will infect bee larvae in their early stages, eventually killing them, and in advanced cases the infected brood give off a foul odor that is very noticeable and vaguely resembles the smell of rotten glue or dead fish. Early in its progression, the disease will turn the pupae into a brownish slime resembling the color of coffee with milk. Over time, the pupa melts away like an ice cream cone on a hot summer day as the bacterium consumes it, causing what's left of the baby bee to ooze down into the lowest part of the cell. The remains of the brood eventually dry out in the bottom of the cell and create a black scale that adheres so tightly to the cell wall that it is difficult for the bees to clean up the mess, especially since the hive's population will be reduced from the action of the disease.

Once a diseased hive has progressed to this stage, the colony is typically so weak that neighboring hives have robbed honey out of the sick hive and spread the foulbrood spores. Compared to healthy brood, whose caps will be slightly raised in shape and light brown in color, the cappings covering the cells containing infants infected with foulbrood will often be dark brown to almost black in color, will have a greasy appearance and a sunken shape, and may be pock-marked with pin-sized holes. Given that brood infected with active *P. larvae* spores fail to hatch, diseased cells can often be found adjacent to healthy brood or surrounded by cells that have been abandoned by healthy hatchlings. As a result, frames of brood in a colony suffering from AFB often take on a spotty pattern, rather than the solid pattern typical of healthy hives where the queen has laid eggs of similar age all adjacent to one another.

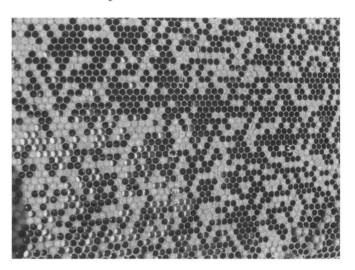

AFB can also be recognized in the field by testing the stringiness of the dead pupae, which are moist and slimy in the early stages after death. When a small twig, piece of straw, or toothpick is inserted into the pupal mass, it will tend to stick to the probe and stretch a quarter to a half inch or so as the probe is removed, much like the sticky, ropelike viscosity of mucus. Dead pupae unaffected by foulbrood disease will not stretch or become threadlike after being poked. American foulbrood is easily spread, and it can destroy an entire apiary if not attended to quickly. Thus it is vitally important that every beekeeper become familiar with and be able to recognize the symptoms of AFB. (See a summary of symptoms in the "Bee Hive Autopsy Results" sidebar on page 61.)

Brood affected by AFB will tend to have dark, sunken cappings (often with pinholes in them) and may imitate a shotgun pattern, with numerous empty cells scattered among the capped cells. Photo by Steve Parise.

P. larvae is difficult to control because in its resting stage the bacterium forms spores that are reported to live for as long as fifty years or more, until favorable conditions allow it to bloom.[2] AFB is most often spread from colony to colony by bees from a neighboring colony robbing out a dead infected hive and carrying the disease back to their own nest, or by drifters that visit from hives infected with the disease. Beekeepers may also unwittingly spread AFB by interchanging frames between sick and healthy colonies, by not cleaning up diseased equipment from a colony that has died from AFB before it gets robbed out, or by feeding colonies honey taken from diseased hives and contaminated with AFB spores. Mixing the honey from a questionable source with water to create a syrup and then boiling the mixture for twenty to thirty minutes will kill any spores that may be present and make the feed safe for use. Luckily, neither AFB nor any other honey bee disease will have a physiological effect on people who handle the bees or equipment, eat the honey, pollen, or comb, or handle or use the beeswax or beeswax products from an infected colony.

Black scales like these on the lower side of the cell are indicative of American foulbrood in its advanced stage. Photo by Steve Parise.

For many years in the United States, the only approved treatment for AFB was the use of the antibiotic oxytetracycline, known commercially as Terramycin. The drug is applied either as the active ingredient in a grease patty or mixed with confectioners' sugar and sprinkled near the brood nest within the hive. However, the patties were not always consumed in a consistent and rapid manner, which would create a situation in which an incomplete dose of the antibiotic would be administered to the hive. The use of such patties has been blamed for the recent revelations of Terramycin-resistant forms of AFB now infecting hives. As a result, the USDA has approved a new and more powerful antibiotic, Tylan, for use against foulbrood disease. This new antibiotic is extremely stable, and rather than being used prophylactically, as oxytetracycline was, proponents say it should be applied only when signs of foulbrood are clearly visible, and it should be applied as a powder shaken into the hive, to reduce the chances that *P. larvae* will develop resistance to Tylan as well. (Antibiotics when applied in powder form should never be administered directly into the brood area, or they may kill the developing larvae and pupae.)

No matter how careful one is, however, antibiotic treatments (when they work at all) simply mask the presence of the disease, preventing viable spores from growing and reproducing. Such treatments do not remove the spores from the hive. As a result, hives treated with antibiotics tend to quickly show signs of reinfection with the active form of AFB once antibiotic use is discontinued. In addition, special care must be taken to ensure that honey being harvested does not become contaminated with anti-biotics. The regular consumption of minute doses of antibiotics in the human diet has been linked to the development of antibi-otic-resistant microbes that cause numer-ous human diseases that resist conventional antibiotic treatments.

To permanently remove American foul-brood from a colony, all AFB spores must be eliminated from the hive. With a strong dose of determination and an investment of time and resources, this can be accom-plished within a single season by removing every bit of drawn comb and honey from an infected colony. Such an endeavor is best conducted early in the season, so that the hive has the maximum amount of time to build new comb and store adequate supplies of honey and pollen for winter. Also, by taking away a colony's drawn comb early on in its yearly cycle, before egg laying reaches its peak, a minimum amount of brood is sacrificed for the greater good.

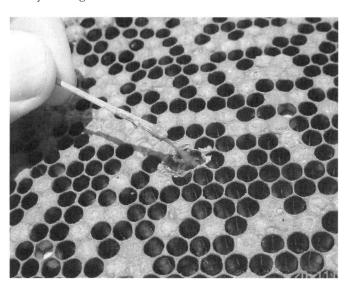

This test for foulbrood can be conducted only on recently deceased larvae that have not yet dried out and formed a black scale on the lower side of the cell. Photo by Steve Parise.

The actual process of eliminating AFB from a hive consists of shak-ing each frame of bees from the hive into a new hive body filled with foundation. If the queen is spotted, it is best to transfer her into her new home by grabbing her wing and placing her on the top bars of the foundation frames so she can crawl down between the frames. (To reduce the chance of an injury that could affect her ability to lay eggs, a queen should never be held by any part of her body other than her wing.) Included in the new hive body of foundation should be one empty frame of drawn comb. This frame of comb is left so that the bees have a place to deposit honey they may have engorged them-selves with during the process of smoking and shaking them into their new home. The very next day, the frame of drawn comb containing these honey deposits should be removed and replaced with a frame of foundation. The colony should then be fed uninfected honey or

191

sugar syrup so that the hive has something to get started on and won't starve should it immediately experience a week of cold, wet weather and not have the opportunity to forage.

By following this procedure, all the beeswax combs and honey that may possibly contain AFB spores are removed from the hive. All of the frames and comb from the infected colonies should be burned, and all hive bodies, supers, inner covers, and bottom boards should be scraped clean of burr comb, which should also be burned. The woodenware itself can be scorched with a propane torch or other device that will heat the surface of the equipment to a high enough temperature to ensure that any remaining AFB spores are destroyed.

> Bees are naturally able to resist . . . disease organisms when they are vigorous and well nourished. The bees also make use of propolis, the most potent naturally occurring antibiotic there is, by shellacking the interior surfaces of the colony with a thin layer of the material to help maintain a sterile environment within the hive.

When I first started beekeeping, I would use a dusting of Terramycin prophylactically in the spring about a month prior to the first major honey flow and again in the fall following the harvest. Over time, while making the effort to use antibiotics less often, I would treat a hive only after it showed symptoms of disease. Unfortunately, because the spores had become embedded in my combs, AFB would consistently pop up in my colonies when left untreated. I knew I needed to change my approach if I was going to get myself off the antibiotic treadmill and keep bees organically. I became motivated to experiment with an operation that I first read about in the book *Elimination of American Foulbrood without the Use of Drugs* by Mark Goodwin and Cliff Van Eaton, published by the National Beekeepers Association of New Zealand.[2] This book gave me the courage to believe that elimination of foulbrood without chemicals was indeed possible and could be successfully accomplished with my hives.

I removed all the drawn combs from my colonies at the end of April, right before the daytime temperatures typically become consistently warm (above 50°F/10°C) and the dandelions begin to bloom, providing the first major honey flow for the season in my area. I gave all the hives a half gallon of sugar syrup to get them started until the dandelion flow kicked into high gear. I was concerned about whether the bees would be able to gather enough honey and build enough comb to supply themselves for the winter, and I was grateful that, during this particular season, we had a very good honey year. Not only did all the hives draw out their foundation and pack away plenty of honey stores for the winter, but enough surplus honey was produced that I was able to harvest a respectable crop that year. That was the year my honey crop was exceptionally light in color, a condition I attribute in part to the fact that all the honey had been stored by the bees in clean, new, lemon yellow combs. That was over four years ago, and I have been blessed with AFB-free bees ever since, without having to once resort to antibiotic use. I am now extremely careful never to allow any questionable frames or equipment to become mixed in among my hives, and I am vigilant in working to prevent someone else's apiary nearby from spreading the disease to my colonies. One of the biggest sources of disease infection and reinfection are the neglected carcasses of dead hives from hobby beekeepers close by.

The keys to maintaining foulbrood-free colonies are regular hive inspections and careful examination of every colony that dies for signs of disease. Once your hives are disease free, regularly removing a couple of the old, darkened combs from the colonies each year and replacing them with frames of foundation will serve to reduce the opportunities for disease spores to build up within a colony's drawn comb reserves. This regular rotation of old combs out of the hive will also provide a steady source of beeswax to be rendered. By the same token, one should make it a practice to avoid frames of comb when purchasing equipment. Unless you know that hard chemicals and antibiotics have not been used on the colony and you have a high level of confidence in the source, accepting used frames is tantamount to taking on someone else's problems.

Should I ever discover another outbreak of foulbrood, I will immediately shake the surviving bees onto foundation and once again destroy all the hive's combs and honey. If this option is not immediately available though, I will move the diseased hive to an isolated quarantine yard until it is possible to do so. It is important to eliminate all combs that exhibit AFB symptoms from your bee yard as soon as possible to avoid having other colonies nearby become infected by the disease spores.

Chalkbrood

Caused by the fungus *Ascosphaera apis*, chalkbrood (CB) disease attacks both worker and drone larvae and pupae. The cottony, threadlike mycelium of this fungus will entomb the developing brood and form a white mass that will harden and mummify the larvae, causing them to resemble small pieces of chalk—hence the name. These mummies turn black as the fungus matures and approaches its reproductive stage. As a result, both white and black mummies can be found in hives with advanced cases of the disease. The chalklike mummies will tend to collect on the bottom board, or by the entrance as house bees clean them out of their birthing cells-turned-coffins. Chalkbrood appears to be a stress-related disease that will weaken the hive but rarely kill it outright. It's somewhat analogous to the common cold. As with a cold, which can be overcome with quality nourishment like chicken soup, CB will typically go away on its own following a good strong flow of honey. Similar to people, a beehive's natural immunity is strengthened when the bees are well nourished with high-quality food.

Incidences of chalkbrood tend to correlate with high moisture and humidity levels within a colony. This suggests that increased hive ventilation may help to prevent this disease. As is the case for many maladies of the honey bee, propagating strains of bees that are resistant to chalkbrood and keeping colonies vigorous and strong are the best ways to prevent infection.

Nosema

Nosema disease is caused by a microscopic organism (*Nosema apis*) that destroys the lining of the honey bee's gut, affecting the bee's nutrition and shortening its life span. Although the only sure way to diagnose nosema is to examine the bee's gut under a microscope, some signs of nosema infection may be observed in the bee yard, including fecal matter within the hive or excessive droppings by the hive entrance. However, dysentery may also cause these signs and be confused with nosema. This occurred in my area of Addison County, Vermont, one year during an abnormally dry summer when the bees collected a lot of honeydew that produced a dark honey with a taste that resembled that of brown sugar. Although not a problem in itself, the indigestible matter in the richly colored honey that accounted for its dark complexion caused severe cases of dysentery within the hives that particular winter, which was a hard one for the bees, with long cold periods that kept them cooped up with few opportunities to go on cleansing flights.

It is said that removing the bee's head and pulling out its guts for inspection can provide a fairly accurate diagnosis of a honey bee suspected of suffering from nosema. The color of a healthy bee's gut will be pinkish or brownish red, whereas a diseased gut will have a dull gray or whitish color. Here in the Northeast, nosema is a concern primarily during the winter months, when the bees may be confined for long periods without the opportunity to relieve themselves of the indigestible matter that builds up in their digestive systems. While the conventional approach includes the chemotherapy antibiotic treatment fumagillin (Fumidil B), the organic beekeeper has no treatment options other than to rear resistant stock and work to keep the immunity level of the hives strong by ensuring proper nutrition, ventilation, et cetera. One way to help reduce instances of nosema is to avoid heavy fall feeding of syrup as this has been correlated with a higher incidence of the disease.[3]

Since I began keeping bees in 1992, I have never used fumagillin or anything else to specifically treat nosema, and neither do most of the beekeepers I know. For some reason, nosema disease, which apparently can be a serious problem in some areas of the United States, has not been much of an issue in my neck of the woods. Whether all the susceptible hives around me have died off, leaving only nosema-resistant colonies remaining, or the threat to colonies from this disease has been overblown, or I have just been lucky, I cannot say. My experience certainly indicates that, even without approved medications for nosema, the naturally inclined beekeeper can certainly raise strong, healthy bees and need not be overly concerned about this disease.

Notes

1. A. I. Root Co., *The ABC & XYZ of Bee Culture* (Medina, OH: A.I. Root Co., 1990): 119.

2. Mark Goodwin and Cliff Van Eaton, *Elimination of American Foulbrood without the Use of Drugs: A Practical Manual for Beekeepers* (Napier, New Zealand: National Beekeepers' Association of New Zealand, 1999).

3. Dewey Caron, from comments made during his presentation at the Vermont Beekeepers Association's winter meeting in Barre, Vermont, January 23, 2007.

CHAPTER 10

THE HONEY HARVEST

Give me a day
That is balmy and shimmering;
Give me a field
Where the clover blooms sway;
O let me linger
Where nectar is gathering,
Deep in the blossoms of flowering May.

Give me a shelter
Where sunlight is filtering,
Tempering music
Of humming and whirr,
Then I will gather
A treasure so glistening,
Fragrant as incense, and precious as myrrh.

—Florence Holt Davison

Of all creatures great and small, the wondrous little honey bee is the only one I am aware of—aside from humans—that consistently takes significantly more from nature than it needs. As long as there is nectar available, good flying weather, and room in the hive to store and ripen the nectar into honey, the worker bees will continue to gather nectar, even after they have put away a far greater supply than is required to see the colony through the winter months. This lack of moderation puts the hive in a vulnerable position. What happens to all that extra honey? Your friendly neighborhood beekeeper comes along and takes the excess honey that the colony worked so hard to gather, process, and store. As a result the bees can teach us to check our greed and not take more than we truly need from the world around us, lest someone is tempted to help him- or herself to our excess in similar fashion.

> Adding another honey super also maximizes the hive's honey-gathering potential by helping to reduce the overcrowded conditions that will cause a hive to cut back on its foraging activities and decide to swarm.

Although apiculturists have little control over the weather and the ability of nearby flowers to produce nectar, we can easily modify the amount of space within a hive for honey storage. To obtain a large honey harvest, the beekeeper must add honey supers to the hive at the appropriate time. The bees will start to fill the brood nest with honey if additional supers are not added to the colony once its combs have all been filled. This is simply because once a hive is full, the only room left is created when a fully matured honey bee chews its way out of its birthing cell. When a hive infringes on the brood nest for honey storage it is said to be "honey-bound," a condition that will typically trigger a colony's swarming instinct. The timely addition of honey supers on a hive helps prevent the queen from becoming honey-bound and the workers from building an abundance of burr comb in their efforts to create more storage space. Adding another honey super also maximizes the hive's honey-gathering potential by helping to reduce the overcrowded conditions that will cause a hive to cut back on its foraging activities and decide to swarm.

"Many hands make light work" is a maxim personified by the colony that keeps its workforce at full strength. The opportunity to add empty supers to a hive before the bees have completely filled the supers they have is just one of the many reasons why conscientious beekeepers conduct regular inspections of their hives. This regular inspection goes hand in hand with an understanding of just how prolific honey bees can be, to ensure that supering is done appropriately and in a timely manner, thus reducing the bees' swarming tendency.

Notice that I talk about *reducing* the tendency to swarm, not eliminating it. The swarm is the hive's natural means of reproduction, thereby increasing the number of colonies in the world. It is a healthy and necessary process for the bees despite what a swarm may mean in economic terms for the beekeeper, and so it is not wise for beekeepers to do everything in their power to prevent swarming. In fact, if your hives are swarming, it is an indication that you are taking pretty good care of them and your bees are healthy and thriving; struggling, weak colonies do not have the resources to issue swarms. Because of our tendency to see everything in purely human terms, the literature all too often refers to swarming in a negative light. From the bee's perspective, however, the swarm is a wonderful thing. Like the birth of a baby, I would invite you to consider the swarm as the miracle of life that it is and celebrate what the world has gained from such an occurrence rather than dreading it and dwelling on what may be perceived as a personal loss.

> Because of our tendency to see everything in purely human terms, the literature all too often refers to swarming in a negative light. From the bee's perspective, however, the swarm is a wonderful thing.

When supering hives, it is generally preferable to utilize drawn comb and use foundation only when the honey flow is heavy. It is also important to note that strong hives will have an easier time drawing out foundation in a timely manner than weak hives. This is not to imply that one should not place drawn combs on a strong hive during a heavy honey flow, or that foundation should never be placed on a weaker colony. It's just that the bees are reluctant to draw out foundation when the honey is only trickling in. So if foundation needs to be drawn, waiting until a good honey flow is in process is the best policy; and the more bees available to do the work, the faster it will get done. In fact, workers are more likely to chew away the edges of the foundation when little or no honey is being made, and they will often deposit the small amount of nectar that is gathered in burr comb rather than begin drawing out foundation.

The hive body typically consists of ten frames of comb, in order to provide the queen with the maximum amount of egg-laying room, but honey supers should contain only eight frames. By using eight frames in a honey super that is made to hold ten frames, the amount of space for the bees to store honey is maximized. At first glance, this might seem counterintuitive. After all, wouldn't nine or ten frames of comb hold more honey than eight? Unlike frames of brood comb, which tend to be consistently built to a certain width, the bees will modify the thickness of the combs they intend to store honey in and build them to fill the available space. Thus, because there are fewer pieces of wood and metal taking up space within an eight-frame honey super, as compared to a ten-frame super, there is more room for the bees to build comb and store honey. Because of the importance of maintaining the hive's bee space, eight is the minimum number of frames that should be used in equipment built for ten frames.

Maximize your honey storage by minimizing the number of frames per super while at the same time maintaining the bee space.

Photo by Alice Eckles.

Given that the modern hive body and super are designed to hold ten frames, it is important to evenly space the frames of comb within the cavity of the box when using only eight or nine frames in order to preserve the bee space. Commercial frame spacers can be purchased, but my preference is to use a thumb or finger as my spacing tool to position the frames uniformly. To accomplish this, first space eight or nine frames out equally in a super or hive body. Then see which of your thumbs or fingers fits between each frame. You may find that you will need to hold your thumb or finger at a certain angle for it to gently touch the frame resting on either side of the space. By inserting the same digit in the same manner between each frame when spacing future supers, adequately spaced frames can be achieved in short order every time. The choice is yours. You can either purchase premade frame spacers or simply use the tools you have on hand—literally! You can then enjoy the added benefit of never losing track of where your frame-spacing tools are, always having them "handy" when you need them.

When it comes to harvesting honey, the organic philosophy and most organic standards dictate that only surplus honey be harvested, and that care must be taken to preserve the integrity of the harvest. The practice of taking almost all of a colony's honey and then feeding the bees sugar or corn syrup in return, although logical from a macroeconomic viewpoint, is neither sustainable nor in the best interest of the bees, since it takes the honey bees' time, energy, and resources to turn nectar, sugar water, or corn syrup into honey. The fact that most corn syrup will contain at least some genetically modified material further complicates this issue. As a result, ecologically minded beekeepers must leave an adequate supply of honey on the hives to provide for the bees' winter needs, and this amount will differ from region to region.

Why spend money on a frame spacer when you have a finger or thumb that will accomplish the same job?
Photo by Alice Eckles.

Once the decision has been made to harvest a honey crop, the organic beekeeper has three basic options available for removing the bees from the honey supers: manual shaking and brushing; blowing; or using some type of bee escape. Shaking and brushing requires little in the way of specialized equipment. A soft-bristled brush is all that is required. The downside of this low-tech approach is that the bees are likely to get riled up unless they are smoked before and after being brushed or shaken from their combs. Care must be taken, however, because too much smoke can adversely affect the flavor of the honey extracted. However, because of its simplicity, this method is a good option for the beekeeper with few hives.

Blowing the bees off the frames of honey to be harvested is a quicker method and is more desirable for those beekeepers with numerous hives. Blower systems are sold by bee supply companies and are similar to leaf blowers, or shop vacuums with the ability to run in reverse. Air is then directed between the frames of the honey super as it sits up on end and the bees are blown through the super and out the other side. Although it's fast and efficient, this method definitely makes the bees angry, and a veil is standard operating equipment whenever one is harvesting with a blower. The wrath of the bees can be tempered somewhat by utilizing the blower during times when temperatures are below 60°F (16°C).

The bee escape is a small device that acts as a one-way door, and it is placed in the center opening of the standard inner cover that features an oval hole in its center. Supers of honey that are to be harvested but

201

are filled with bees are placed above the inner cover, and as bees move down toward the brood area over a period of days, they pass through the bee escape, which prevents them from reentering the honey super above later on. The escape board works on the same principle and takes the place of an inner cover fitted with a bee escape. The larger opening on the escape board allows the removal of bees from honey supers to proceed at a faster rate than with the bee escape, typically in twenty-four to forty-eight hours. The downside of using either the bee escape or the escape board is that it requires a minimum of two trips to the hive to bring in the harvest. In addition, bees are reluctant to leave supers that contain even the smallest amounts of brood, and these frames will require shaking or brushing anyway. Bee escapes and escape boards work best when evening temperatures drop down into the fifties or below, encouraging the bees residing in the honey supers above to leave and gravitate toward the brood nest cluster. When using a bee escape or escape board, it is critical that all other openings—between supers, in the inner or outer cover, and so on—that are large enough to allow bees to exit and enter are sealed up. Scraps of beeswax, mud, drywall compound, or duct tape all work well in this regard.

> Drawn comb represents a substantial investment on the part of the hive in terms of the amount of time, labor, and resources. It is estimated that worker bees must consume anywhere from seven to eight pounds of honey for every single pound of beeswax that they excrete through their wax glands.

A key to a successful honey harvest is to wait until the bees have transformed the nectar into ripe honey with a moisture content below about 18 percent. A thin layer of wax is the covering the bees will use to seal combs filled with fully ripened honey. Once capped, the honey in these cells is protected from contamination from dirt and debris, thievery by robbers, and, to some extent, moisture. If frames of honey that contain primarily uncapped cells are harvested, the honey extracted will typically contain too much moisture and it will ferment over time. The general rule of thumb is to be sure that, at minimum, 75 percent of the combs that are extracted are fully capped. This means that on average each frame can contain up to 25 percent of its cells uncapped, or that for every frame of uncapped honey that is harvested, three frames of fully sealed combs must be extracted along with it. Those who want to remove the guesswork should purchase a refractometer, a precision instrument that will accurately measure the moisture level in honey.

Honey Processing

In the old days, harvested honeycomb was squeezed to separate the honey from the wax. Today, extracting machines efficiently remove honey using centrifugal force in much the same way that water is forced out of clothing in our washing machines. Removal of honey without destroying the wax allows the bees to reuse the frames of honeycomb. The comb is the furniture that makes up the honey bees' home. The bees are born in these combs, live on them throughout their life, and use them for food storage. Drawn comb represents a substantial investment on the part of the hive in terms of time, labor, and resources. It is estimated that worker bees must consume anywhere from seven to eight pounds of honey for every single pound of beeswax that they excrete through their wax glands. Aside from sacrificing some of the honey harvest to allow combs to be built, the beekeeper's investment lies in the cost of acquiring foundation and inserting it into each frame.

Before frames of ripe honey sealed neatly within the comb can be extracted, the wax cappings must first be removed so that the honey contained in each cell can flow out freely. Organic production does not allow for the use of heat; approved ways of uncapping frames of honey range from the use of machines that mechanically uncap each frame to manual methods such as the uncapping knife and the capping scratcher or uncapping fork. An uncapping knife with a serrated edge is the best way to effectively remove cappings by hand and without heat. The handle of the uncapping knife should be offset and the blade long enough so that each end of the knife will ride along the bottom and top bar of each frame, helping to guide the cut and prevent gouging into the delicate combs. Care must be taken to hold the knife at the appropriate angle so that the blade will not cut into the wooden frame components. The scratcher comes in handy for removing cappings from the areas of comb that the bees do not draw out past the plane created by the outer edges of the wooden parts of each frame.

The decision as to whether to use deep supers, shallow supers, or medium supers to collect and harvest the honey is a personal one that each beekeeper has to make for him- or herself. In terms of strict efficiency, the choice is simple. Each shallow frame holds less honey than a medium or deep frame, and therefore it will take longer to extract the same amount of honey when using shallow supers for honey collection than it will with the frames of medium or deep supers.

On the other hand, the larger the super, the heavier it will be when filled by the bees with the ambrosial gifts of their labor. In recent years, bee supply companies have promoted smaller boxes that hold a maximum of eight frames as an alternative for those who wish to save their backs. Whether you choose to utilize eight-frame or ten-frame hive bodies and supers, I strongly recommend that once you have made your choice you stick with one type or the other. Otherwise, the day will inevitably come when you find yourself needing a hive body or super for an eight-frame hive, and the only empty equipment you have available to use is made for ten frames, or vice versa.

When extracting honey, variables such as temperature and the types of materials that are allowed to come in contact with the honey become factors. As with all good production practices, uncapping, extracting, and bottling equipment must be clean and free of foreign materials and debris prior to use. Stainless steel, food-grade plastic, and other food-grade materials must be utilized to reduce the possibility of compromising the harvest's integrity. In keeping with this train of thought, all organic standards limit the temperature that honey is allowed to be exposed to, with most specifying 95° to 120°F (35° to 49°C) as the maximum range, and some going even further, also limiting the length of time that honey may be held at the maximum temperature. This is because the enzymes in honey are adversely affected at temperatures approaching 120°, and high temperatures will affect the honey's color and flavor. These temperature requirements refer to the temperature range of uncapping equipment and bottling equipment. Conventional honey that has been heated will stay liquid on the shelf for a much longer period than unheated honey before crystallizing, and because it is clear, it is typically filtered to remove the tiny particles of pollen, propolis, and beeswax that may have entered the honey during the extraction process. Not only do these minute particles detract from the clarity of the final product, but they also act as nuclei around which crystals may form, speeding up the crystallization process and shortening the shelf life of the liquid honey. Organic honey standards prohibit the filtration of honey, thereby preserving the nutritional benefits that the bits of pollen, beeswax, and propolis impart to the final product. As a result, most natural and organic honey that is available will be crystallized, with the exception being honey that has been freshly harvested and honey from floral sources that naturally resist crystallization, such as Tupelo honey.

To remove the majority of the beeswax, pollen, and propolis that gets into the honey during extraction without using heat, a screen or sieve can be used. However, running honey through a sieve can be a slow process, and when harvesting large amounts of honey, settling tanks are the preferred method of making the final product presentable. As honey sits, foreign debris in the honey will slowly rise to the surface, where it will float. The relatively clear honey at the bottom of the tank can then be drawn off and bottled.

Once the harvest has been bottled, the national organic standards have much to say about the label that goes on the final product. The NOP separates organic foods into three categories: 100 percent organic, organic, and made with organic ingredients. The labeling requirements for each are as follows:

(a) Products sold, labeled, or represented as "100 percent organic." A raw or processed agricultural product sold, labeled, or represented as "100 percent organic" must contain (by weight or fluid volume, excluding water and salt) 100 percent organically produced ingredients. If labeled as organically produced, such product must be labeled pursuant to § 205.303.

(b) Products sold, labeled, or represented as "organic." A raw or processed agricultural product sold, labeled, or represented as "organic" must contain (by weight or fluid volume, excluding water and salt) not less than 95 percent organically produced raw or processed agricultural products. Any remaining product ingredients must be organically produced, unless not commercially available in organic form, or must be nonagricultural substances or nonorganically produced agricultural products produced consistent with the National List in subpart G of this part. If labeled as organically produced, such product must be labeled pursuant to § 205.303.

(c) Products sold, labeled, or represented as "made with organic (specific ingredients or food group(s))." Multi-ingredient agricultural product sold, labeled, or represented as "made with organic (specified ingredients or food groups(s))" must contain (by weight or fluid volume, excluding water and salt) at least 70 percent organically produced ingredients which are produced and handled pursuant to requirements in subpart C of this part. No ingredients may be produced using prohibited practices specified in paragraphs (1), (2), and (3) of § 205.301(f). Nonorganic ingredients may be produced without regard to paragraphs (4), (5), (6), and (7) of § 205.301(f). If labeled as containing organically produced ingredients or food groups, such product must be labeled pursuant to § 205.304.[1]

The USDA allows the green and white organic logo to be used only on products consisting of 95 to 100 percent organic ingredients. Unfortunately, because the controversial issues surrounding honey have not yet been resolved, resulting in the NOP dragging its feet on the establishment of organic honey standards, the use of the USDA organic logo on honey labels has itself become controversial. Some organic certification agencies—typically those that had organic honey standards in place before the NOP standards were released—will allow the use of the term *organic* as long the apiculturist meets the requirements as written in their standards. Unfortunately, the many challenges that are faced by the beekeeping industry are spurring a rapid pace of advancement in beekeeping methods and technologies and such standards are typically outdated and in dire need of being re-written. Other agencies that don't have standards in place have tended to put the work of writing standards on hold, the rationale being that it makes little sense to spend time and effort creating a set of standards for honey when the NOP regulations, once released, will trump their own. When national standards will finally be realized is an open question. Not only is there currently little momentum behind the creation of national organic honey standards, but the controversial issues surrounding the establishment of such standards are so sticky that they suck in the participants and mire down progress, like a bee drowning in a tank full of honey. As with all things, honesty is the best policy when it comes to organic agriculture. The penalties for fraudulently labeling organic products consist of a civil penalty of ten thousand dollars for each occurrence, in addition to criminal penalties under federal law.

Honey Inventories

One of the most important elements for a colony's survival during winter is access to adequate stores of honey to ensure that the hive does not starve before spring rolls around. Beekeepers who are truly concerned with the welfare of the bees will harvest only the excess honey that has accumulated during the course of the year, and they are careful to leave the colony with enough honey to see it through the winter. The amount of honey a hive requires for its winter use will differ from location to location. Here in Vermont, bees typically need a minimum of sixty to eighty pounds of honey per colony. A good rule of thumb is to harvest all the supers that are filled with honey with the exception of the last full super that is situated just above the brood within the hive. During the harvest, as each full super is lifted off a hive, the super directly below it should be inspected for brood. If there is a significant amount of brood mixed in with the honey stored in the super below, leave the super that is chock-full of honey and sitting above it in place. In addition, make sure that any nucleus colony created during the course of the year that was unable to fill at least a shallow super above the brood nest with honey receives a full super harvested from a stronger hive. This way we ensure that every colony goes into winter with a full super of honey above the brood nest. Capped honey above the brood nest is key to the colony's winter survival. During the course of the winter months the cluster will naturally make its way up through the hive, often ending up near or even right up against the inner cover by spring. Thus, the anatomy of a hive prepared for winter finds the bees clustered around the brood nest near the bottom of the hive with plenty of honey stored above to provide room for their slow but steady upward winter migration.

This standard operating procedure goes hand in hand with the management technique that utilizes hives consisting of a deep (nine and five-eighths inches) hive body sandwiched between two shallow (five and eleven-sixteenths inches) supers, which act as shallow hive bodies, as opposed to the conventional hive consisting of two deep hive bodies. There are a number of benefits to using two shallows and a single deep as the main body of a hive. The greatest benefit is realized when having to feed colonies. When it comes to feeding a hive, the best food we can offer is honey already sealed in the comb. As long as care has been taken to ensure that honey fed to a hive does not contain American foulbrood or other disease organisms, a full sealed super of honey is first-class bee feed. A weakened hive makes an especially good recipient of such feed, because it will save that hive's bees from expending their energy filling combs and converting sugar syrup into honey. When using shallow or medium supers as part of the main hive unit, honey can be fed in the comb more efficiently than if only deep hive bodies are used for the hive proper. A nuc colony that has filled its deep hive body with honey, pollen, and brood can typically be provided with all the feed it will require for the winter by adding a full shallow containing about thirty pounds of honey on top of the deep hive body, rather than giving the colony another full-depth super that can hold as much as seventy-five or eighty pounds of honey—especially since there is honey already stored along with the brood in the hive body below.

Hives that feature a deep hive body sandwiched between two shallows allow for greater efficiency when feeding full supers of sealed honey.

The other big plus, of course, is the fact that shallows full of honey are much easier to lift than full deeps—a benefit that becomes more and more significant the older one gets. Also, if mice get into the hive during the winter, they typically build their nest in the bottom super. In such instances it is preferable to have shallow frames of drawn comb damaged by mice rather than the more valuable deep frames.

If no frames filled with honey are available, extracted honey can be fed to needy colonies. Because of the threat of disease, all extracted honey from questionable sources should always be mixed with some water and the resulting syrup boiled for twenty minutes to a half hour before being fed to a hive, so that any disease spores that the honey contains will be destroyed. However, boiled honey is not as desirable as clean raw honey that has been unheated, because heat destroys the enzymes naturally found in Mother Nature's sweetener. If honey is not available, the next best feed to use is white sugar. The reason white sugar is preferred over brown sugar, turbinado sugar, or other types when making syrup for bees is that white sugar has had all its nonsugar components removed and will not contribute to dysentery within the hive during the winter. The same refining process that provides empty calories that prove detrimental for human consumption is a little less harmful when used for bees. A ratio of one part sugar to one part water is recommended for spring feeding, whereas a two-to-one ratio is best when mixing up sugar syrup for feed in the fall, a time when hives need to make and store as much honey as possible as quickly as possible. The use of hot water will help ensure that all the sugar will get completely dissolved—just make sure the syrup cools sufficiently before feeding it to the bees. Honey produced from sugar syrup is not as desirable as honey produced naturally, because syrup is deficient in the natural minerals and enzymes found in honey produced from fresh nectar.

Most organic standards require that organic sugar be used when making sugar syrup, and they restrict the use of high-fructose corn syrup (HFCS). This is because feed made from HFCS will result in honey that is likely to contain genetically engineered material, which has the potential to weaken the vitality of the hive, making it more prone to contracting diseases or becoming a target for parasites. However, it is most likely that should certified organic HFCS become available, organic standards would find it acceptable for use as bee feed. Should this occur, beekeepers concerned about the well-being of their hive should avoid HFCS anyway. This is because corn syrup contains two types of sugars that are poisonous to bees; stachyose and raffinose. As a result, the life span of bees is reduced when they are fed HFCS. Additional toxic compounds are formed if the syrup is exposed to high temperatures before being fed to hungry colonies. Such "bad batches" of HFCS will cause honey bees to die even sooner than those that receive untainted corn syrup.[2]

When fall feeding is complete, each hive should be given a "heft test" by grabbing the back of the hive by the handhold of the bottom or middle hive body and lifting. If the hive feels like it's filled with lead and you are unable to move it without excessive strain, then the hive is in good shape for winter as far as honey goes. It takes some experience to consistently estimate the amount of honey stored in a hive accurately by hefting alone. Initially it is a good idea to open up a few hives to confirm the level of capped honey they contain and to get a good sense for how heavy a hive should feel when properly provisioned for winter. One final manipulation that should be made prior to winter is to position the center of the brood nest so that it is located near the middle of the hive body, with an equal number of honey frames on either side. This way the colony will have equal access to nourishment no matter which direction they may move during the course of the coming months.

Although winter in the southern United States is not marked by the extreme cold that characterizes the North Country, it is often still characterized by a dearth of nectar sources. Lack of sufficient amounts of stored honey is as serious a problem for the southern beekeeper as it is for those in the North. Depending on location, a colony heading into winter in the South may require almost as much honey as one in the North, and unless it is well supplied, brood rearing will be held in check, the result being a colony that is weak and unfit to fully take advantage of the first major flows of honey come spring.

Notes

1. *Vermont Organic Farmers 2004 Certification Guidelines and Applicant Information,* Richmond, VT (2004): 70–71, www.nofavt.org.

2. Bob Harrison, "Weslaco Bee Lab and Current Research," *American Bee Journal* 147 (4): 323–326.

CHAPTER 11

ORGANICS AND THE EVOLUTION OF BEEKEEPING

She imbibes the song of joy evoked by sunlight on petals.

—The Bee Mistress as quoted in
The Shamanic Way of The Bee by Simon Buxton

Agriculture is the backbone of any sustainable economy. By combining the gifts of Earth, Air, Fire, and Water, agricultural activity is one of the few forms of production that creates new wealth for society, wealth that did not previously exist in any form. Most economic activities other than agriculture simply add value to previously created wealth or materials, to create a profit. Historically, major economies have always experienced an agricultural revolution before making a transition to an industrial model.

For well over a decade now, the American market for organic foods has been the fastest growing segment of the U.S. food industry, achieving gains of 20 percent or more every year, and by 2007, organic food is projected to become an industry worth $30.7 billion, with no end of growth in sight. In 1990, Congress recognized this growing demand for organic products and passed the Organic Foods Production Act (OFPA). This act called for the creation of a National Organic Standards Board (NOSB), a citizen review panel whose job is to develop a National List of approved substances allowed for use in organic production and to advise the U.S. secretary of agriculture on program implementation policies. Prior to the OFPA, organic standards were developed through the years by numerous certification agencies, all working on their own, and as a result, organic standards had become anything but standard. Due to the wide diversity of certification organizations, many different regulations governing the production of organic crops were developed. Thus, a major part of the reasoning behind the establishment of federal organic standards was the desire to standardize the rules governing organic certification so that they were consistent from state to state and region to region.

Transitioning to Organic Apiculture

Although the OFPA was passed in 1990, it wasn't until 2002 that a final set of organic regulations was finally produced. Despite going through a number of amendments and modifications since then, the NOP has yet to specifically address organic honey production. The patchwork of organic standards that are currently being maintained by various certification agencies throughout the United States will sooner or later be replaced by the national standards, once they have been developed by the NOSB and codified by the USDA. A review of current organic apiculture standards indicates that there will probably be five basic components with which the organic beekeeper will have to contend once the NOSB completes its work.

The first issue to grapple with will be documentation. Everything that goes into or comes out of organic hives will need to be documented. This means that each hive will have to be given a unique identification name or number to differentiate it from all the other colonies in the apiary. If heat is used during any part of the processing and bottling operation, it will have to be monitored and documented. It has always been a good idea to keep a journal of observations, manipulations, treatments, and honey harvests anyway, but organic standards *require* the apiculturist to take this practice to the next level. Once we get into the habit of keeping careful records of each hive, it is easier to track performance over long periods, and this allows us to more easily identify the breeding stock with the highest potential.

Forage type and availability is the second major component of organic honey bee management. As indicated earlier, it will be challenging, to say the least, to locate bee yards in areas where only organically raised crops and unsprayed wildflowers that conform to organic standards are found. Unless future federal standards make a radical shift from what has historically been the case, organic regulations will mostly limit organic apiculture to wilderness areas and to very large organic farms, where an apiary can be established in the middle of the farm to prevent bees from potentially foraging on blossoms growing on plants that have been sprayed with synthetic pesticides or grown with chemical fertilizers.

Much has been said, and continues to be said, regarding the necessity of beekeeping and honey bees for the proper pollination of many crop varieties. The standard line that is repeated is that approximately one-third of all cultivated food crops require the presence of bees for farmers to realize the sizable harvests they have come to rely upon, and the world's hungry have come to need. Eliminate honey bees and the pollination services they provide, and the economic and social upheaval that would result from the massive loss of yields among affected crops would be staggering. What is not often discussed, however, is the important role that agriculture plays in sustaining the beekeeping industry.

> Eliminate honey bees and the pollination services they provide, and the economic and social upheaval that would result from the massive loss of yields among affected crops would be staggering. What is not often discussed, however, is the important role that agriculture plays in sustaining the beekeeping industry.

The Champlain Valley that encompasses Lake Champlain and lies between New York's Adirondack Mountains and Vemont's Green Mountains is a vast, fertile expanse that supports a strong agricultural industry made up primarily of dairy farms. These farms maintain many acres of clover and alfalfa to feed their cattle, and it is these same fields that make the Champlain Valley such a wonderful place to keep bees. Beekeeping in this little corner of the world relies upon dairy farmers for its viability. Eliminate the farms and forests would soon replace the fields, reducing the availability of bee forage so drastically that the number of honey bee colonies the valley would be able to support would likely be reduced by an estimated 60 to 70 percent. Thus the interdependence that exists between the world of the honey bee and the plant kingdom directly impacts the relationship between farmers and beekeepers. It is this dependence that is the most limiting factor for certified organic honey production.

Until there is a significant increase in the percentage of total farm acreage in the United States that is certified organic, the limited availability of forage for certified organic hives will severely depress the potential number of organic colonies that can be managed. This means that organic honey will continue to be rare and producible only by those lucky enough to have access to the limited number of areas where certified organic production can take place. Because organic agriculture makes up only about 2 to 3 percent of the overall agricultural acreage in the United States, the majority of organic apiaries will have to be located in and around wooded areas, parks, and other wild places that typically do not provide an abundance of nectar sources and thus do not contribute to large honey surpluses and bumper honey crops. These circumstances will force the organic bee industry to remain small and localized well into the foreseeable future.

> The health of a hive is the sum total of a wide variety of large and small details. Such a state of health can be accomplished consistently not by approaching the craft of beekeeping through the industrial, one-size-fits-all model, but only by recognizing and respecting apiculture as the natural biological activity that it is.

However, in many cases apiculturists can adopt and benefit from another key factor of organic production: hive management. Different techniques and materials will have to be found to replace the hard chemicals upon which most beekeepers have become dependent. The one-size-fits-all silver-bullet approach will no longer suffice due to variations in scale, goals, time limitations, beekeeping skill, and climate. Each beekeeper will have to develop his or her own unique style that will utilize the three categories of disease and pest control measures: mechanical (screened bottom boards, traps, secluded locations, burning, et cetera), biological (beneficial fungi, *Bacillus thuringiensis*, genetically tolerant queens, et cetera); and soft chemicals (essential oils, organic acids, et cetera). In order to add value to their honey and differentiate their finished products from the rest of the pack, savvy marketers will use as a selling point the fact that alternative methods eliminate the use of toxins and antibiotics.

All aspects of harvesting, processing, and storing of hive products will come under additional scrutiny. Bees will have to be separated from their honey by mechanical means rather than through the use of chemical fume boards. Old galvanized extractors will fall by the wayside as modern stainless-steel and food-grade processing equipment becomes the order of the day. The beekeeping industry is fortunate that its primary product, honey, will not support the growth of any bacteria, fungi, or mold that is harmful to humans. Honey, in fact, is so safe and stable that samples of honey found in ancient Egyptian tombs were still edible. Additionally, the often repeated myth that babies should not be given any honey due to the chance that they may contract botulism is severely suspect. I have yet to come across a single documented case of infant mortality due to the feeding of pure, natural honey. When added to a petri dish containing a botulism colony, honey will kill the botulism growing there. Honey's natural qualities—as an antimicrobial agent and preservative—have thus covered up for many a careless operation throughout the years, but such sloppiness will no longer be tolerated in the world of organic certification.

Because most standards in effect today do not allow the organic beekeeper to just let a hive die, the use of conventional methods to restore colony health are typically required whenever a hive is in jeopardy. A "hospital yard" will have to be established, with procedures for dealing with sick hives that must be pulled out of organic production. Beekeepers will also have to take measures to prevent any possibility of commingling organic with nonorganic honey.

Another component that will change will be the labeling and marketing of the final product. All organic labels must by law state the name of the USDA-approved third-party certification agency responsible for confirming the organic integrity of the finished product. Given that the folks responsible for your certification will have their name on your label, they will want to ensure that you meet all specifications laid out in the standards. Labeling details such as font size and the positioning of certain statements (for example, the name of the certification agency must appear directly under or next to your company contact information) will become paramount. Organic beekeepers will want to have their organic certification agency preapprove their label before sending it off to the printer, or they run the risk of having to modify and reprint labels after the fact when they don't meet with official approval.

A sixth and final factor should be mentioned here: the significant investment that beekeepers will need to make to certify their honey as organic. Not only will there be costs associated with the trial and error of transitioning to organically harmonious ways of keeping bees, but there will also be a yearly fee that will have to be paid to cover the cost of third-party inspections. That's right. Your apiaries, your honey house, your books—everything—will be scrutinized to make sure you have met all organic requirements; so be sure to do your homework before going for your certification, and maintain rigorous adherence to the standards. The fees are typically determined by the size of your business as measured by yearly sales figures. An exemption has been made for small operations that gross less than five thousand dollars a year. These small businesses don't have to pay to be able to call their honey organic, but they still have to meet all the requirements of the standards. Such small producers do not have to undergo third-party scrutiny and are thus taken at their word. In the end, though, they are not allowed to use the organic logo on their label, nor do they receive a certificate of organic authenticity for their products—a document that more and more retailers are asking for these days to meet their customers' expectations.

Only products that have received USDA-approved third-party inspection can sport the USDA organic logo.

Once you've established your plans and set up your systems to meet the organic standards, the work of organic beekeeping can begin. First you have to transition your bees into organic management. Since it is far easier to manipulate the interior of a beehive than a farm of many acres, the transition period for honey bees, depending on the standards followed, tends to be less than a year. With all the challenges of getting organically certified with honey bees, this is one of the few bright spots; for instance, dairy livestock are required to pass through a full one-year period of organic management before being certified, and agricultural land must undergo a three-year transition period before becoming fully organic. Of course, all bets are off once the national standards are finally in place, because they may call for a very different transition period.

One might logically assume that those beekeepers who have managed to get certified through one of the sets of standards written and enforced by an individual certification agency prior to the establishment of the NOP organic apiculture regulations will have an easier time making the transition to national standards. Unfortunately, this is not necessarily so. The NOP defines honey bees as livestock, and as such, they are subject to all the rules governing organically raised livestock. For example, organic livestock may be fed only 100 percent organic feed. The OFPA, however, takes the position of previous legal opinions that distinguish between honey bees and traditional livestock, whose foraging activities can be more easily controlled. As a result, the question of whether bees are livestock and must be fed exclusively organic feed while under organic management is about as clear as mud. Until there is an answer to this question issued by the NOSB and codified in the NOP, individual certifiers are being left to their own devices to sort out the issue of honey bees' status as livestock.

Some certification agencies allow the use of nonorganic supplemental feeding and the use of the USDA organic logo on the honey label. Other agencies don't allow one or the other, and in some cases, neither practice among beekeepers is certifiable under their standards. Agencies that do not allow use of the USDA organic logo on honey have more leeway in interpreting the standards, since the final product falls under their own organic standards rather than the national regulations. Many of these certification groups were active before the implementation of the NOP and have their own organic logos that beekeepers can use on their honey label, as long as they meet the relevant group's standards.

The State of Organics

The organic movement was started as an alternative to the status quo and "business as usual." It helped to create the not-for-profit-only business model in which quality is more important than quantity, and now organic has proven so successful as a concept that big business is taking notice and wants a piece of the action. Needless to say, all the confusion and indecision surrounding aspects of the NOP serve the interests of conventional mass-market businesses well. Producers that want to take advantage of the marketing strength of an organic label without having to change their conventional practices have benefited the most from the confusion. This has resulted in the organic standards becoming subject to political manipulation, an influence that does not always have the integrity of the regulations in mind. Some companies have even been able to "beat the system" by lobbying their congressional representatives to pass legislation that bends the rules and allows exemptions to the 100 percent organic feed for livestock rule, or lets chickens or cows be certified organic even if they don't get to graze outside for any significant time in their lives. Because of this shifting playing field, where every interest is jockeying for advantage, individuals and organizations that wish to preserve the integrity of the federal organic standards must be vigilant in defending them against efforts that would weaken them in the eyes of the consumer.

The "good old days," when local producers sat around a table and hashed out the issues until they all came to a clear conclusion that everybody could live with, must seem like a real blessing to today's certification administrators, who must contend with so much uncertainty. The bureaucratic nightmare created by the federalization of organic agriculture is a major hurdle for the conscientious organic certifier. In addition, the delays and foot-dragging create additional challenges and difficulties for certifiers and producers. Within the first few years of implementation of the NOP, the USDA has made and proposed changes to the standards only to backpedal later, after the consuming public expressed widespread opposition. The USDA has also released statements indicating that it has taken a particular position on an issue only to stall on releasing the official written decision. This apparent indecision creates an air of insecurity among both certification agencies and producers. True organic production requires a degree of certainty that in many ways is missing from the National Organic Program as it has been implemented in its early years. One can only hope that this situation will improve with time.

Whether consciously or unconsciously, many beekeepers, bee researchers, and beekeeping organizations are helping to perpetuate the air of indecision and confusion surrounding the finalization and implementation of organic apiculture standards. This group includes a variety of interests, from beekeepers who think that all apiculture is natural and that there is "no such thing as organic beekeeping" to agricultural schools that promote a chemical mind-set that tends to overlook natural alternatives and that are influenced by research money doled out by chemical/agribusiness companies. Many of these same institutions and individuals will express the feeling that anyone who is claiming that he or she is keeping bees organically is lying and must be cheating somehow. This viewpoint is understandable and to be expected given the high level of pesticide and antibiotic "misuse" that occurs among conventional beekeepers. Far too many beekeepers do not use beekeeping chemicals and drugs as directed by the label or use unapproved pesticides and antibiotics in their hives. Although these activities are defined as federal crimes under the law, weak enforcement and lack of testing provide the opportunity for rampant abuse. Thus, because cheating is all too common among conventional beekeepers, it stands to reason—the thinking goes—that organic beekeepers must be cheating as well. However, unlike conventional beekeeping, there is a third party looking over the shoulder of organic producers to ensure that they do not cut corners and cheat the system. This is one of the reasons why the consumer is willing to pay a premium for organically certified foods.

> Honey bees and the people who care for them will be an essential component in the development of local and regional food systems, which will be required in the future to provide for our long-term food security and maintain a sound, sustainable basis for economic activity.

It's a good thing consumers have not yet gotten wind of the corner-cutting activities of some conventional beekeepers. It would take only one well-publicized incident of chemical or antibiotic contamination to ruin the sterling reputation that honey enjoys as a pure, natural product. The consumers' best defense in this regard is to establish a personal relationship with their local beekeeper, ask a lot of questions, and determine for themselves whether the production practices of their local supplier meets with their approval and philosophical preferences. It is precisely this desire by many consumers that is part of the driving force behind the current expansion of the organic food industry, as well as the increasing number of farmers' markets and community-supported agriculture (CSA) farms that eliminate the middleman and allow farmers to deal directly with the end user. Beekeepers can substantially increase their profit margins and gain significant control over their distribution by packaging their honey themselves and marketing directly to wholesale distributors, retailers, and consumers, rather than supplying the bulk commodity market.

Another reason for the resistance to organic apiculture by some beekeepers is that organic honey is perceived as a threat. Not only do these folks believe, either correctly or mistakenly, that they will be unable to meet the standards for organic production, but they worry they will have to answer to customers who will want to know why their honey is not certified organic. The fear is that nonorganic, conventionally produced honey will be viewed as inferior to organic honey. This is disturbing for many of these beekeepers, since they may sincerely be trying to use chemicals and antibiotics appropriately and are doing their best to produce high-quality products. Many of these folks think that meeting the tough requirements for organic production is simply not possible, because they have such high colony numbers, are unaware of the organic management practices they could be utilizing, or are established in a location that would preclude the option of organic certification due to the close proximity of nonorganic foraging sources. When faced with questions from their customers, many beekeepers will be faced with the choice of trying to explain their use of toxic chemicals and antibiotics in the hive or simply lying. As a result, many in this struggling industry—those who are unable to take advantage of the marketing power of the organic label—will find the business climate all the more challenging as organic apiculture gains in popularity.

Unfortunately, the industrial agricultural model emphasizes profit and relegates stewardship of the land to secondary status at best. The organic philosophy, on the other hand, places stewardship of the land and its denizens as a more important goal than the guarantee of a bumper crop. Systems and processes that renew and rejuvenate themselves define organic agriculture that is truly sustainable in the long run. By investing in the health of the hive and following the Hippocratic oath of "First, do no harm," the organic apiculturist profits by working in the honey bee's best interests, rather than profiting through exploitation. The health of a hive is the sum total of a wide variety of large and small details. Such a state of health can be accomplished consistently not by approaching the craft of beekeeping through the industrial, one-size-fits-all model, but only by recognizing and respecting apiculture as the natural biological activity that it is— an endeavor that touches on the health and well-being of the land and all of its inhabitants.

The Future of Organic Beekeeping

Honey bees and the people who care for them will be an essential component in the development of local and regional food systems, which will be required in the future to provide for our long-term food security and maintain a sound, sustainable basis for economic activity. The reliance on toxic, petrochemical-based pesticides by large agribusinesses in their attempt to "feed the world" has resulted in the contamination of our soil and water and has led to the serious decline of natural pollinators in many areas. Their corporate message is based on the assumption that we must accept this environmental degradation in return for an end to hunger, despite the fact that world hunger is an issue of distribution and geopolitics, not one of production.

Our industrial model of farming has created many farms that consist of acres and acres of a single crop. The use of pesticides on these fields has eliminated most of the natural pollinators, and as a result, farmers have become totally reliant on the services of migratory beekeepers. As the supply of native colonies for providing pollination services continues to decrease in the face of pests and disease, the costs of renting honey bees will climb. The agricultural industry that already has had to rely on importing bees from other states must now import bees from other countries. The economic and ecological costs associated with transporting hives on tractor-trailer trucks across country, or by plane from overseas, to provide pollination services for large farms and orchards will continue to rise significantly unless alternative, sustainable, and ecologically sound fuel sources can be fully developed. Unfortunately, in the United States, government and industry seem more intent on trying to hang on to what's left of the world's declining oil supplies than investing in alternatives that will take us into the future. My sense is that factors such as these will ultimately result in a return to smaller farms being run on a human scale and a new reliance on local apiaries and beekeeping outfits for pollination. These farms will produce a diverse variety of crops to feed the families that work them along with the citizens in the surrounding communities.

As part of this shift, more and more farms will elect to keep their own hives for pollination purposes in answer to both the increasing costs associated with renting hives and the challenge created by

221

the dwindling supply of bees available for hire. This will result in the improvement of the overall health of the honey bees, because they will not have to undergo the stress of transportation, risk exposure to disease and parasitic pests in other areas of the country, or suffer from the limitations in diet that occur when hives are transported long distances and placed in the middle of acres and acres of a single crop during the time of bloom. The bees will also benefit from the reduction in overmanipulation and artificial stimulation with pollen substitutes and sugar or corn syrup that many migratory beekeepers utilize in their effort to equalize and build up their hives prior to performing pollination services. We say that hives that have been fed in such a manner are stronger because the feeding stimulates brood production, but what does this artificial feeding do to the bee's immune system and the overall immunity of the hive?

> By taking back control and responsibility for supplying their own inputs, processing, bottling, and marketing, beekeepers can become much more profitable, and their operations will be more likely to successfully weather times of economic stress.

I envision beekeeping following the path that all of agriculture will have to travel to become more profitable and sustainable in the future: small-scale producers will market products locally in their final form at wholesale or retail prices, and they will take advantage of the increased profit margins that can be obtained through value-added products, as opposed to producing a low-margin bulk commodity. By taking back control and responsibility for supplying their own inputs, processing, bottling, and marketing, beekeepers can become much more profitable, and their operations will be more likely to successfully weather times of economic stress. By reducing the reliance on outside inputs and developing management techniques that increase soil fertility and livestock health over time rather than depleting them, organic agriculture offers real hope for the future of farming. Beekeeping and beekeepers, by necessity, must be an integral part of this agricultural shift.

One of the most obvious and contentious issues that impedes progress on national organic honey standards is the establishment of appropriate buffer zones between organically managed hives and things that would compromise organic integrity. The proliferation of the products of genetic engineering throughout the United States alone may make organic honey and pollen production a moot point, depending on the level of GMO contamination that is allowed in products from the hive. However, the quagmire of certification issues travels much deeper. For instance, should organic farms and orchards be required to use certified organic bees for pollination? Such a requirement might make sense when we look at the established precedents that have already been set throughout the organic regulatory process. For example, nonorganic ingredients may not be used in products labeled as organic, or 100 percent organic, if an organic option is available, and synthetic inputs are allowed only if they appear on the National List of approved and regulated synthetics. What will it mean if standards provide for the pollination of blossoms on organic plants by bees that have been exposed to chemicals unapproved for organic use? When we recognize the incredibly large numbers of hives required for agricultural pollination, and the shortages that already exist in the number of conventional hives available to pollinate nonorganic crops, an allowance is clearly going to have to be made for the use of nonorganic hives pollinating organic crops—at least until the number of certified hives increases dramatically.

Such a spectacular growth in the number of colonies that are managed without toxic chemicals and drugs is a realistically attainable goal. If I can keep my bees healthy and productive without resorting to toxic chemotherapy treatments, each of you reading this book can do so, too. As you try out different approaches to keeping bees, it is wise to evaluate each new management protocol on just a few colonies initially to determine how well it will work in your situation. Once you have grown comfortable with the various methods and techniques with which you are experimenting, they can be expanded to the rest of your operation.

The honey bee is an endless source of fascination, variation, and contemplation. So much so that the claims that bees have been written about and studied more than any other insect on Earth are not at all far-fetched. The fact that the little honey bee manifests such incredible healing power in the world—healing that works on many levels, from plants to humans—is a real blessing. Unfortunately, both the huge managed-colony losses and the disappearance of feral swarms, which has been consistently reported in many areas of the globe during the past decade, indicate that the honey bee—like every single major ecological system and many of the species on our planet—is under profound stress.

The answer, in my opinion, will not be found in a fast and easy magic-bullet solution. Instead, it will require a combination of methods and a shift in our current management techniques that must be utilized with success over the course of many decades, and without developing newly resistant pests or diseases. By adopting a variety of natural, nontoxic approaches to hive management, like some of the ones described in this book, visionary beekeepers will help move their industry toward a permanent solution—or at the very least help buy more time for the honey bee. Such creativity, focused on helping the beekeeping industry to progress in a sustainable manner, has the potential to carry the ancient craft of keeping bees well into the future.

AFTERWORD

As this book was going through its final editing phases and being pre-pared for the printer, word started traveling around beekeeping circles about the alarming and sudden loss of tens of thousands of managed bee colonies. This phenomenon is characterized by a rapid collapse in which the vast majority of the adult bees disappear from the hive. Hives that seem to be fine one day are, two to three weeks later, found dead and relatively empty. No significant numbers of dead adult bees are found in or around the collapsed colonies, as would normally be the case. To add to the mystery, these hives typically have capped brood and plenty of honey and pollen stored away in the combs. Strangely enough, these honey stores remain unmolested by neigh-boring bees that would normally rob out such bounty with little delay. Even more interesting is the fact that there is a noticeable interval of up to two weeks or so before the usual parasites (wax moths and small hive beetles) move in and feed off the carcasses of the formerly bus-tling hives.

No area in the United States seems to be immune to the ravages of this phenomenon, with reported colony collapses stretching from California to the East Coast, encompassing some thirteen states so far.[1] This significant blow to the beekeeping world is poorly understood and threatens both honey production and the pollination industry in the United States. All of this is leaving some people to openly wonder whether the U.S. beekeeping industry can survive. Some states, like Pennsylvania, now have less than 50 percent of the number of managed colonies that they had just twenty-five years ago.[2]

Researchers are beside themselves trying to figure out what exactly is going on. Dead bees from affected hives are being collected and dissected. Samples of comb, brood, honey, and pollen are also being studied. Surviving hives from affected apiaries have been transferred to research centers for observation and testing. Surveys are being carried out among affected and nonaffected beekeepers in an effort to evaluate the various management practices in use. All this in the hope of providing a glimpse into the cause, or causes, behind this disturbing event.

Members of the Pennsylvania and Florida departments of agriculture, along with researchers from Penn State University, have prepared a preliminary report that focuses on this situation. According to their initial findings, cited above, some common threads have started to reveal themselves as a result of these investigations. The beekeeping operations that have been impacted tend to be migratory in nature. As previously mentioned in this book, moving hives is stressful on the bees. First you have the stress of the move itself. Such factors as confinement, wide temperature swings, and the potential interruption of egg laying by the queen all lead to apiary losses, with 10 percent loss common among hives that have been transported long distances.[3] Due to the pressure migratory operations are under to provide colonies with the high population numbers that are demanded by pollination contracts, such hives are typically fed carbohydrates (sugar) and protein (pollen substitutes) in order to stimulate brood rearing. While this practice increases honey bee numbers within the colonies, such unnatural food sources add to the stress on the bees' digestive systems and ultimately can compromise and weaken the immune systems of each individual within the hive.

Another interesting condition that was found to be common among the affected beekeepers interviewed was that all their operations had suffered "some form of extraordinary stress" at least nine weeks before the initial hive deaths were observed.[4] The exact nature of this stress varied widely: overcrowded apiaries, forced crop pollination on forage with minimal nutritional value, and a serious dearth of nectar and pollen, all three of which in turn lead to nutritional deficiencies. Additional stresses include drought or contaminated water supplies and the damaging effects of varroa mites. The one common factor was that all these conditions adversely affect the immunity of the bees, making them all the more susceptible to viruses and other disease organisms.

Added to all this information is the fact that every operation affected had used antibiotics in an effort to keep the bees healthy, and the vast majority had used a miticide at some point during the 2006 season.[5] The temporary support that such chemical use offers the hive, combined with the sublethal buildup within the combs of toxic components of various compounds, may have further compromised the natural immunity of the colony. Initial investigations of the numerous samples collected from these dead colonies have revealed a variety of pathogens within the hives and within the bees themselves. As a result, it does not appear that any single disease can be attributed to these occurrences so far, and so the new epidemic has been named colony collapse disorder (CCD).[6]

From the natural beekeeping perspective, CCD sounds suspiciously like the results of the industrialization of beekeeping coming home to roost. The constant strain on hives that have been repeatedly pushed to their limits is bound to take its toll sooner or later. It is interesting to note that nowhere in the preliminary investigative report, or from anecdotal sources, is there any hint so far that those beekeepers practicing natural or organic apicultural techniques have been targeted by this horrible condition.

I suspect that at least part of what we are observing here is the result of the fouling of the interior of the hive's environment from the toxic pollutants and pesticides that are brought back to the hive by foraging bees, to the slow buildup of the numerous chemical agents used to control parasitic activity. The use, and abuse, of synthetic chemotherapy mite controls is helping to create conditions wherein the combs are no longer healthy places for bees to make a home. The key clue that points to this possibility is the reluctance of other bees and parasites to move in and scavenge over the remains of these dead colonies. Animals intuitively know which foods are good for them and which are not.

This situation reminds me of the anecdotal reports from farmers who notice that their livestock, when given the choice between conventionally grown grain and grain grown from genetically engineered (GE) seed stock, will shun the GE feed in favor of the conventional food every time. Such animals are observed to eat GE grain only when they are given no other choice. By leaving a stick of butter and a stick of margarine next to each other outside, we can observe a similar phenomenon. Birds, insects, and other creatures will devour the butter, which is full of vitamins, minerals, protein, and fats, and will ignore the margarine, which is composed of hydrogenated oils and trans fatty acids and is generally so lacking in nutrition that it takes artificial coloring, multimillion-dollar advertising efforts, and a multipronged disinformation campaign to fool the consuming public into eating the stuff and thinking it's good for them. Could it be that the bees are telling us, in a similarly unambiguous way, that the methods we are using to create hives full of bees, brood, honey, and bee bread, although well intentioned, are actually condemning the bees to death? By pushing our hives to be economic engines for honey production and pollination services, are we forgetting to observe the basic necessities the bees require in order to fully live out their lives in a healthy and productive manner?

Granted, major losses of bees have been experienced by beekeepers in the past. The "Isle of Wight disease" that first was documented on the eponymous island in the United Kingdom, and that was responsible for major losses of managed colonies over a relatively short period of time, turned out in the end to be caused by the tracheal mite. Perhaps the current epidemic will prove to have some identifiable and relatively simple cause after all.

One potential silver lining among this whole grisly CCD situation is that our political leaders and the general public may finally understand the terrible importance of the honey bee in supporting our system of food production. If these reported colony deaths continue to escalate, the adverse effect from the loss of hundreds of thousands of managed colonies and their pollination activities will ripple throughout the agricultural community. When growers are unable to secure adequate pollination for their crops, production and quality will drop precipitously, and food shortages—along with skyrocketing prices—are sure to follow.

Beekeepers have never had much political clout. However, once our voices are combined with those of other producers of agricultural commodities, whose welfare is tied to ours, the resulting chorus should be loud enough to rise above the din of competing interests. Once the plight of beekeepers starts to have an impact on the average person's pocketbook and daily life, politicians will be forced to take more notice and greater resources finally will be diverted to addressing the challenges of the beekeeping industry. The question remains, however: Will these resources be used to provide temporary fixes—"Band-Aid" solutions that try to maintain our business-as-usual approach—or will they be used to help blaze a new path toward a more sustainable future for the beekeeping industry? Time will tell.

Notes

1. Dewey Caron and Mike Palmer, remarks made during presentations to the Vermont Beekeepers Association, Barre, Vermont, Tuesday, January 23, 2007.

2. Dennis van Engelsdorp, Diana Cox Foster, Nancy Ostiguy, Maryann Frazier, and Jerry Hayes, "'Fall-Dwindle Disease': Investigations into the Causes of Sudden and Alarming Colony Losses Experienced by Beekeepers in the Fall of 2006. A Preliminary Report" (December 15, 2006), and Dewey Caron and Mike Palmer, presentation, January 23, 2007.

3. Ibid.

4. Ibid.

5. Ibid.

6. Ibid.

Glossary

abscond—when an entire colony of honey bees abandons a hive and settles in a new location

AFB—abbreviation for American foulbrood

Africanized honey bee—the strain of honey bee that is descended from the African bees brought to Brazil in the 1950s. The Africanized honey bee migrated up through Central America and has spread into the southern United States. This bee is characterized by its aggressive defensive behavior that is a result of its historical exposure to an environment that provided short and sporadic honey flows coupled with a large number of natural predators.

AHB—abbreviation for Africanized honey bee

American Bee Journal—monthly periodical on beekeeping published by Dadant & Sons, Hamilton, Ill.

American foulbrood—bacterial disease that is highly contagious among honey bee colonies, produces a foul odor in advanced stages, and usually leads to the death of a colony unless treated with antibiotics or the burning of infected combs

apiary—location where one or more beehives are kept

apiculture—things of or relating to beekeeping

apiculturist—one who practices apiculture

apitherapy—use of honey bee and hive products (honey, propolis, pollen, royal jelly, and bee venom) for health and the treatment of disease conditions

artificial swarm—see *nucleus colony*

Bacillus thuringiensis—bacterium that, when ingested by certain insect larvae, produces crystals and spores that kill the host

bee bread—pollen that has been slightly fermented by the bees and stored in comb cells, usually near the brood nest

Bee Culture—monthly publication on beekeeping published by the A. I. Root Co. of Medina, Ohio; formerly called *Gleanings in Bee Culture*

bee escape—device that fits into the inner cover of a hive and allows for the one-way passage of honey bees; used primarily for removing bees from honey supers prior to harvest

bee glue—see *propolis*

bee space—the area between objects in a hive that measures approximately ⅜ of an inch and within which the bees tend not to build comb, preferring to keep the space free and open for use as a passageway within the hive

beeswax—a substance produced by the wax-secreting glands in young honey bees' abdomens that is molded and shaped by worker bees into combs

birthing cell—the cell within a comb into which the queen lays an egg and the resulting bee emerges once it has developed to the point where it is ready to join the rest of the colony as a mature bee

bottom bar—the lower or bottom piece that makes up a modern beekeeping frame

bottom board—the piece of hive equipment that makes up the bottom or floor of the modern beehive

brace comb—wax comb built between frames within a hive

brood—young honey bee larvae, pupae, and immature bees

brood chamber—the area where the queen bee lays eggs and the hive raises brood

Bt—abbreviation for *Bacillus thuringiensis*

burr comb—the wax comb built on the top, bottom, and ends of frames within a hive

cappings—the thin wax covering bees build over cells to cover them. Cells of honey and developing brood of a certain age are capped in such a way. The cappings used to seal cells full of honey must be removed as part of the modern honey extraction process.

cappings scratcher—see *uncapping fork*

chalkbrood—a disease state created by the presence of the fungus *Ascosphaera apis* that affects honey bee larvae and pupae, resulting in their "mummification" and taking on the appearance of little pieces of chalk

chilled brood—brood that is killed due to exposure to cold temperatures

cluster—the formation of honey bees within a hive that occurs in cold weather of about 55°F (12°C) or below and characteristically consists of a dense ball of bees with the queen and brood at its center

colony—a family of honey bees occupying a single cavity

colony collapse disorder—a term developed during the winter of 2007 to describe the death of honey bee colonies that exhibit the following symptoms: The rapid decline of a hive's population to the point where few or no adult bees are found dead or alive in the hive, either on the combs or on the bottom board. If any living bees are present they are only the queen accompanied by a handful of very young workers. Brood in various stages are present in larger quantities than any remaining cluster can maintain. There is a noted delay of 2–3 weeks after the initial collapse before there is any robbing/scavenging activity by other bees, wax moths, small hive beetles, wasps, or hornets within the dead hive.

comb—a structure manufactured by honey bees out of beeswax and consisting of hexagon-shaped cells fitted side by side and used by bees to raise brood and store honey and pollen; also called honeycomb

comb honey—honey in the comb prepared for human consumption as stored by honey bees and found naturally occurring within the hive

dearth—a lack or absence. In beekeeping, *dearth* typically refers to the lack of nectar or pollen available for forging honey bees.

deep super—boxes 9⅝ inches tall that are filled with frames of comb and used to provide room for a colony to store excess honey intended for harvest by the beekeeper. When used to house the brood nest of a hive, it is called a hive body.

division board feeder—type of feeding device that is shaped like a frame and takes the place of one or more frames in the hive, thereby providing feed close to the brood nest

drawn comb—comb that has been built up from a sheet of foundation

drone—the male of the honey bee species characterized by a wider body mass and larger eyes as compared to the female honey bee. The drone is not endowed with a stinger and does not participate in any of the work activities taken up by the hive.

drone brood—the immature stage of the male honey bee

drone layer—an infertile queen bee that is capable only of laying primarily unfertilized eggs that develop into drones

economic injury level—the level of varroa mites within a hive that will negatively effect the the hive's honey production to the extent that it will adversely impact the income of beekeepers who rely on honey sales for some portion of their income. This level is typically considered to be around 3,000 mites.

EIL—abbreviation for economic injury level

end bars—the two pieces of a frame that fit between the top bar and the bottom bar

escape board—a flat device shaped like an inner cover that is placed between two honey supers and allows for the one-way passage of honey bees; used primarily for removing bees from honey supers prior to harvest

European foulbrood—a bacterial disease that is highly contagious and is similar to American foulbrood in that it usually leads to the death of a colony unless treated with antibiotics or the burning of infected combs

forage range—the distance covered by worker bees searching for items to bring back to the hive

foraging—the gathering activities of the worker bee. Honey bees typically forage for nectar, pollen, plant resins to make propolis, and water.

foundation—a thin sheet, typically made of wax or plastic, that has the outline of hexagonal-shaped honeycomb cells embossed onto both sides of its surface

frame—a man-made contraption typically made of wood or plastic consisting of a top bar, a bottom bar, and two end bars, and usually fitted with foundation to encourage bees to build comb within the confines of the frame. Frames are designed to hold and support honeycomb and allow for easy removal from a hive for inspection and harvesting without damaging the comb.

GE—abbreviation for genetic engineering

gene—a piece of DNA (deoxyribonucleic acid) that carries information for the biosynthesis of a specific product within the cell. A gene is the unit by which inheritable characteristics are transferred to succeeding generations in all living things. Genes are contained within, and arranged along the length of, the chromosome. Each chromosome of a species has a specific number and arrangement of genes, which govern both the structure and the metabolic functions of the cells and thus the entire organism. Genes provide information for the synthesis of enzymes and other proteins and specify when these substances are to be manufactured. A change in either gene number or gene arrangement can result in mutation (a change in inheritable traits).

genetic engineering—the process of altering or manipulating the genetic material (DNA) of cells or organisms to make them capable of making new substances or performing new functions. This gene-splicing technique transports selected genes from one species to a different, often totally unrelated species. These genes or portions of DNA molecules are removed from the donor (insect, plant, animal, or other life-form) and inserted into the genetic material of a virus. The virus is then allowed to infect the recipient organism so that the organism ends up absorbing both the virus and the genetic material. When the virus replicates within the organism, large amounts of the foreign, as well as the viral, material are manufactured.

genetically modified organism—a life-form that is the product of genetic engineering. GMOs are considered to be intellectual property and thus these life-forms can be patented under current laws.

GMOs—abbreviation for genetically modified organisms

hive—a home for bees. Hives manufactured by humans typically consist of a bottom board, an inner cover, an outer cover, and one or more boxes stacked directly one over the other and filled with frames of comb or foundation.

hive body—the body of a hive typically consisting of a box $9\,5/8$ inches deep and containing frames of comb within which can be found the majority of the brood within the hive

hive tool—an essential part of beekeeping equipment consisting of a piece of flattened metal that resembles a minicrowbar and is used for prying, scraping, and manipulating the parts of a hive

honey—the nectar from flower blossoms that has been gathered, processed, and concentrated by honey bees into a sweet, viscous material composed primarily of glucose and fructose, along with numerous trace minerals, and a small amount of vitamins, proteins, and enzymes

honey bound—the state in which a queen bee is unable to lay a significant number of eggs because the cells in the brood area are filled with honey due to the lack of available space in the rest of the hive to store processed nectar

honeycomb—see *comb*

honey flow—when flowers are producing enough nectar to allow significant honey storage that results from a concentrated period of nectar foraging usually over a relatively short period; sometimes referred to as a nectar flow

honey house—a building usually in or near an apiary where beekeeping equipment and supplies are stored, where honey is extracted and stored, and which serves as the beekeeper's workshop

honey super—a box used for honey production that can consist of a shallow, medium, or deep super

inner cover—a cover that sits on top of the uppermost super on a hive, underneath the outer cover, and contains one or more openings to allow for ventilation and the use of feed buckets

laying worker—an infertile female honey bee (worker) that has been stimulated to develop her egg-laying capacity due to the absence of a fertile queen within the hive. The eggs laid by a laying worker are infertile and will result only in the production of drones.

medium super—a box that is $6\,5/8$ inches deep and typically contains frames of comb in which the bees can store honey or raise brood

migratory beekeepers—apiculturists who move hives to one or more locations during a season to pollinate a specific crop or to take advantage of a honey flow

natural comb—honeycomb that is built by bees without the aid of foundation

nectar flow—see *honey flow*

NOP—abbreviation for National Organic Program

NOSB—abbreviation for National Organic Standards Board

nosema—a disease caused by the microorganism *Nosema apis* that affects the midgut of the honey bee

nuc—abbreviation for nucleus colony

nucleus colony—a new hive of bees created by the beekeeper and typically containing three to five frames of bees, brood, honey, and pollen. A nucleus colony is a miniature version of a regular beehive and often is used to expand an apiary or replace dead colonies. Sometimes referred to as an artificial swarm.

OFPA—abbreviation for Organic Foods Production Act of 1990

outer cover—the outermost cover of a modern beehive that protects the hive from precipitation

pollen—dust-sized particles produced by the anthers of plants for the purpose of procreation. Pollen is picked up by the hairs on the honey bee's body and then formed into a compact clump by the bee, who brings it back to the hive, where it is stored in fermented form in honeycomb cells. Pollen is necessary for the rearing of brood.

proboscis—the tongue of the bee

propolis—a sticky substance with powerful antibiotic properties that is composed mainly of plant resins. The bees use propolis both as a building material to anchor parts of the hive and fill in cracks and holes and as a varnish over the interior of the hive to provide a sterile environment and prevent the growth of bacteria and disease.

queen—a completely developed female honey bee that is capable of laying both fertile and unfertile eggs and usually has a life span of approximately five to seven years.

queen cage—a small box containing a screen used to hold and transport a queen bee

rabbet—a recessed area located along the upper inside edge of the ends of supers and hive bodies for the purpose of providing a place for the frame to rest when hanging within the hive

rendering—the act of melting down beeswax, combs, and cappings and removing the impurities

reversing—a hive manipulation that expands the area used for brood rearing. Reversing a hive typically involves moving a hive body full of brood and honey from the top of the hive down to the bottom board and placing the mostly empty hive body that had occupied the space on the bottom board to the top of the hive to provide room for the queen to lay as she naturally moves upward during the course of the season.

robbing—the removal of stored honey from a hive by members of another nest or hive

sac brood—a virus-induced disease of bee brood that is contagious but not too serious

sealed brood—young bees in the pupa stage whose cells have been sealed over with a thin layer of mostly beeswax. Sealed drone brood will jut out from the comb and have a bulletlike shape, whereas sealed worker brood will appear flatter and more in line with the surface of the comb.

shallow super—a box that is 5 $^{11}/_{16}$ inches deep and typically contains frames of comb within which the bees will store excess honey

SHB—abbreviation for small hive beetle

small hive beetle—a parasitic beetle whose larvae feed on the honey, pollen, and brood of weak or dying hives. Feces from the feeding larvae cause honey in combs to ferment. Bees finding themselves in such a situation will tend to abscond.

smoker—a metal apparatus equipped with bellows into which a burning material is introduced with the intention of producing smoke that can be directed toward honey bees in an effort to render them docile and easy to work with

SMR—abbreviation for suppressed mite reproduction

super—a box filled with frames of comb or foundation that is placed above the brood chamber of a hive to provide the colony with the space to store excess honey typically intended for harvest by the beekeeper

supering—the act of adding supers filled with frames to a hive for the purpose of providing the colony with additional honey storage space

supersedure—the process by which the worker bees raise a new queen from a fertilized egg to replace the mother queen, typically because she is old and failing

suppressed mite reproduction—see *varroa-sensitive hygiene*

swarm—a group of worker bees of various age groups and a queen bee that have left behind a mature colony to start a new one. The swarm will usually leave behind a hive that is full of honey, pollen, brood, worker and drone bees, and a virgin queen that has recently hatched, or is about to hatch, out of her birthing cell.

top bar—the uppermost cross-member of a frame. The ends of the top bar will rest in the rabbets of the supers or hive bodies to suspend the frame and make it accessible to the bees.

tracheal mite—a microscopic mite that lives in the trachea of the honey bee. This mite is very difficult to detect without the aid of a microscope.

uncapping fork—a multipronged instrument used to scratch the cappings off honey sealed in cells prior to its extraction from the comb.

USDA—abbreviation for United States Department of Agriculture

varroa—a mite from Asia that has spread around the world parasitizing European honey bee colonies. Because the European honey bee is not varroa's native host and has not developed methods of naturally keeping the mite's population under control, varroa typically overwhelms European honey bee colonies within one to two years if left untreated, causing colony death primarily from the transmission of viruses from mite to bee.

varroa-sensitive hygiene—a genetic characteristic expressed within a hive of honey bees by the tendency of the workers to identify developing brood that is sharing its cell with actively reproducing varroa mites and to remove the infested brood from the colony, thereby breaking the reproductive cycle of the mite. Formerly known as suppressed mite reproduction.

VSH—abbreviation for varroa-sensitive hygiene

wax moth—a moth that parasitizes weak or dead colonies, laying its egg masses in cracks and crevices within the hive

wax worm—the larvae stage of the wax moth. Wax worms bore their way through combs spinning silken tunnels and eating beeswax combs, pollen, and honey as they go.

worker bee—a female honey bee with an underdeveloped reproductive system that carries out the majority of the work necessary for the proper functioning of a colony and has a life span of approximately five to six weeks during the summer and four to six months during the winter

worker brood—young worker bees in the larval and pupal stages

Resources

The following resources are not meant to be a complete list of beekeeping resources, but are provided to get you started. The following companies and organizations are simply ones that I have had dealings with or am familiar with.

General

A. I. Root Co.
623 W. Liberty St.
Medina, OH 44256
800-289-7668
e-mail: info@BeeCulture.com
Web site: www.BeeCulture.com
Publishers of *Bee Culture Magazine*

American Apitherapy Society
4835 Van Nuys Blvd., #100
Sherman Oaks, CA 91403
818-501-0446
e-mail: aasoffice@apitherapy.org
Web site: www.apitherapy.org
Information on the use of honeybee products for the treatment of disease conditions

Bee Alert Technology, Inc.
1620 Rodgers T., Suite #1
Missoula, MT 59802
406-541-3160
Anti-theft technology

Betterbee
8 Meader Rd.
Greenwich, NY 12834
800-632-3379
 and
Dadant & Sons
51 South 2nd St.
Hamilton, IL, 62341
888-922-1293
e-mail: ABJ@Dadant.com
Web site: www.dadant.com
At the time of this writing, both were sources for small cell foundation of 4.9 millimeters

Better Way Wax Melter Honey
  Processors Ltd.
116 11th Street SE
Altoona, IA 50009
515-967-4952
"The Cyclone" lactic acid vaporizer

Dadant & Sons
51 South 2nd St.
Hamilton, IL, 62341
888-922-1293
e-mail: ABJ@Dadant.com
Web site: www.dadant.com
Publishers of the *American Bee Journal*

Golden Bee Products
4405 Lake Villa
Metairie, LA 70002
504-456-8805
Ventilated, sting-resistant bee suit

Heilyser Technology Ltd.
685 Dalkeith Ave.
Sidney, BC V8L 5G7
Canada
250-656-8727
e-mail: orioleln@shaw.ca
Web site: www.members.shaw.ca/orioleln
Oxalic acid vaporizer

Beekeeping Supply Companies

Betterbee
8 Meader Rd.
Greenwich, NY 12834
800-632-3379
Web site: www.betterbee.com

Brushy Mountain Bee Farm
610 Bethany Church Rd.
Moravian Falls, NC 28654
800-233-7929
e-mail: info@brushymountainbeefarm.com

Dadant & Sons
51 South 2nd St.
Hamilton, IL 62341
888-922-1293
e-mail: dadant@dadant.com
Web site: www.dadant.com

F. W. Jones and Sons
Box 1230
Bedford, Quebec J0J 1A0
Canada
450-248-3323
e-mail: info@fwjones.com

Humble Abodes
636 Coopers Mills Rd.
Windsor, ME 04363
877-423-3269
e-mail: humbleabodes@prexar.com

Mann Lake Ltd.
501 S. 1st St.
Hackensack, MN 56452
800-880-7694
Web site: www.mannlakeltd.com

Maxant Industries
P.O. Box 454
Ayer, MA 01432
978-772-0576
e-mail: sales@maxantindustries.com

Rossman Apiaries
P.O. Box 909
Moultrie, GA 31776
800-333-7677
Web site: www.gabees.com

Walter T. Kelley Co.
P.O. Box 240
807 W. Main St.
Clarkson, KY 42726
800-233-2899
Web site: www.kelleybees.com

Breeders of Bees with Mite-Resistant Genetics

B. Weaver Apiaries
16481 CR 319
Navasota, TX 77968
936-825-7312
e-mail: info@beeweaver.com
Web site: www.beeweaver.com

Champlain Valley Bees and Queens
P.O. Box 381
Middlebury, VT 05753
802-758-2501

Glenn Apiaries
P.O. Box 2737
Fallbrook, CA 92088
760-728-3731
e-mail: queenb95@aol.com
Web site: http://members.aol.com/queenb95

Hardeman Apiaries
906 South Railroad Ext.
P.O. Box 214
Mt. Vernon, GA 30445
912-583-2710

Harper's Honey Farm
421 Louveteau Rd.
Carencro, LA 70520
337-298-6261
e-mail: labeeman@russianbreeder.com

Pendell Apiaries
P.O. Box 40
Stonyford, CA 95979
530-963-3062

Organic Certification Agencies with Apiculture Programs

MOFGA Certification Services, LLC
P.O. Box 170
294 Crosby Brook Rd.
Unity, ME 04988
207-568-4142
e-mail: mofga@mofga.org

Vermont Organic Farmers (VOF)
P.O. Box 697
Bridge St.
Richmond, VT 05477
802-434-4122
e-mail: info@nofavt.org
Web site: www.nofavt.org

Oregon Tilth, Inc.
470 Lancaster Dr. NE
Salem, OR 97301
503-378-0690
e-mail: organic@tilth.org
Web site: www.tilth.org

Organic Crop Improvement Association
(OCIA)
6400 Cornhusker, Suite 125
Lincoln, NE 68507
402-477-2323
e-mail: info@ocia.org
Web site: www.ocia.org

Index

Master Gardening Series

If you enjoyed *Natural Beekeeping* please consider these other great gardening books by Chelsea Green Publishing.

Four-Season Harvest
$24.95 | ISBN 978-1-890132-27-9

The New Organic Grower
$24.95 | ISBN 978-0-930031-75-6

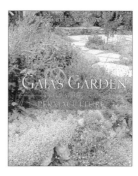

Gaia's Garden
$24.95 | ISBN 978-1-890132-52-1

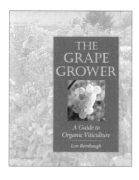

The Grape Grower
$35 | ISBN 978-1-890132-82-8

The Apple Grower
$40 | ISBN 978-1-931498-91-3

The Herbalist's Way
$30 | ISBN 978-1-931498-76-0

Edible Forest Gardens, Vol. 1
$75 | ISBN 978-1-931498-79-1

Edible Forest Gardens, Vol. 2
$75 | ISBN 978-1-931498-80-7

Edible Forest Gardens, Vol. 1 and 2
$150 | ISBN 978-1-890132-60-6

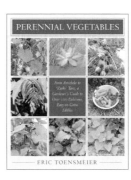

Perennial Vegetables
$35 | ISBN 978-1-931498-40-1

To place an order please visit
www.chelseagreen.com or call (800) 639-4099.

Food & Health

If you enjoyed *Natural Beekeeping* please consider these other
Food and Health books by Chelsea Green Publishing.

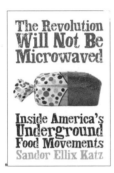

The Revolution Will Not Be Microwaved
$20 | ISBN 978-1-933392-11-0

Full Moon Feast
$25 | ISBN 978-1-933392-00-4

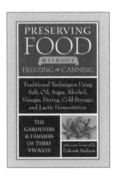

Preserving Food Without Freezing or Canning
$25 | ISBN 978-1-933392-59-2

Wild Fermentation
$25 | ISBN 978-1-931498-23-4

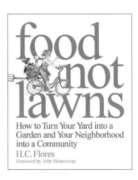

Food Not Lawns
$25 | ISBN 978-1-933392-07-3

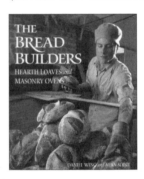

The Bread Builders
$35 | ISBN 978-1-890132-05-7

The Slow Food Guide to Chicago
$20 | ISBN 978-1-931498-61-6

The Slow Food Guide to New York City
$20 | ISBN 978-1-931498-27-2

The Slow Food Guide to San Francisco
$20 | ISBN 978-1-931498-75-3

To place an order please visit
www.chelseagreen.com or call (800) 639-4099.

CHELSEA GREEN PUBLISHING
the politics and practice of sustainable living